the ground at my feet

the ground at my feet

sustaining a family and a forest

ann stinson

Oregon State University Press   Corvallis

Cataloging-in-Publication Data is available from the Library of Congress.

⊗ This paper meets the requirements of ANSI/NISO Z39.48-1992 (Permanence of Paper).

First published in 2021 by Oregon State University Press
Printed in the United States of America

 **Oregon State University**
OSU Press

Oregon State University Press
121 The Valley Library
Corvallis OR 97331-4501
541-737-3166 • fax 541-737-3170
www.osupress.oregonstate.edu

For My Brother
Steven Douglas Stinson

Contents

PART ONE

Outer Bark

The outer bark is the tree's protection from the outside world. Continually renewed from within, it helps keep out moisture in the rain and prevents the tree from losing moisture when the air is dry. It insulates against cold and heat and wards off insect enemies.

I sit on a recently felled Douglas fir and look around. It is early evening and our loggers are gone for the day. Floaty gnats inhabit the slanting light.

A sword fern frond reaches out and touches my notebook. Its other fronds merge with Oregon grape, salal, fir branches, cedar boughs, and chunks of bark. The soil is dark and soft and airy. I touch it. I make three dark stripes on each of my cheeks. The soil is drier than I expect; my stripes are smudges.

I want to roll around in the newly exposed dirt and hear what it has to say. It smells rich and full of secrets. Everything has been opened up. Stories await my hearing. This soil knows about the storms and floods and eruptions. It knows about the Cowlitz peoples, the Hudson's Bay Company trappers, the first White settlers. And does it hear my brother's voice in the raven's call? Does the land remember his love? This forest was my growing ground from nine to eighteen, then a place to visit from Japan and the East Coast, and now home again. How can I better hear what it has to say to me? From my seat on the log, I observe a large old snag in the middle of the cut. We've left it as a perch for hawks and crows. The broken branches stand black against slow-moving clouds backlit by the setting sun. In this gloaming, I listen to the trees.

At the edge of the cut lies Gemini Grove, six acres of majestic hundred-year-old trees my sister-in-law Lou Jean and I have set aside. We go there for walks and contemplation, and the grove shelters critters that like deep shade. It is also home to my brother Steve's Tree, where we go to remember him.

As the harvest enters the third week, Lou Jean and I walk the wetland trail in the grove. In the low spots, water stands at six inches. Last year's alder leaves line the bottom, and new lovage nudges through the surface. A long strand of green-gray moss floats over my blue rubber work boots. Later I look up the moss. Common name: witch's hair lichen; Latin: *Alectoria sarmentosa*. Lou Jean likes to gather it and place it around the ceramic woodswoman that holds some of Steve's ashes. She likes him to be warm.

We have hired a father and son team, Peter Sr. and Peter Jr., to do the logging on this land above the Cowlitz River in southwest Washington. From the trail we can hear the grind and thump of their machines and then can see them beyond our fifty-foot boundary. This logging has daylighted our trail, bringing sun into pockets of fern and old-growth stumps sunken in deep shade for decades. Peter Sr.'s saw whines from across the clear-cut as we enter a darker section of the grove. A tree falls and shakes the ground.

Peter is hand-falling. Our trees are too big for the modern-day processors most of the industry uses. I'm glad. This method is more personal, a more gracious relationship between faller and tree.

As Peter falls them, his son Peter Jr. walks down each tree with a chainsaw, cutting the branches close to the trunk. Next he climbs into a processor, a huge bright orange Doosan that cuts the tree into logs.

With another Doosan machine, a loader, he'll pick up the logs and place them on their truck.

I'm curious about Doosan, a Korean company, and find its website:

> We trace our history to 1896. Our first product was Bakgabun, a women's cosmetic.
> Doosan is made up of two Korean words.
> "Doo" means a unit of grain, while "San" means a mountain.
> Together, they mean "little grains that can build a mighty mountain,"
> suggesting that great things can be achieved
> when even the tiniest forces join together in a unified effort.

Some of our Douglas fir goes to Korea to be used in temples. These logs must be longer than thirty-six feet and larger than thirty-two inches top diameter. I wish all our trees went to such treasured purposes.

Our cedar is milled at Reichert Shake and Fencing, a family-owned mill just a mile and a half from our forest. Other local mills process fir into two-by-fours, plywood, telephone poles, and other utilitarian products. Douglas fir logs with the fewest defects are exported to Japan for home building, while rougher logs we export to China, where some are used to construct concrete forms. At least the imprint of the knotty wood grain stays embedded in the concrete, an echo of forest life from across the Pacific. I begin to imagine traveling to Asia to see and touch our wood in its new home.

A sampling of defects used to grade exported Douglas fir:

AS: Age Stain
BT: Beaver Tail
CF: Cat Face
EK: Excessive Knots
F: Fluted or Flared Butt
FC: Freeze Crack
HC: Heart Check

OH: Off-Center Heart
PR: Pitch Ring
SB: Snow Break
SK: Spike Knot
SP: Spangle
WS: Wind Shake

Another defect, known as a "school marm," is the fork in a treetop log. The current spec sheet for a Longview sawmill reads: "Buck out school marms back to a single heart; no forked tops." As a former schoolteacher, I take offense at this even as I laugh.

We can sell the bucked-out school marms and other rough wood for pulp to make paper. We will keep leftover parts of the trees for firewood. Woodworking friends come to look at the burls, yew, and other "hobby wood."

Logs that are too big are a problem. Any log over thirty-three inches diameter on the large end is deemed "oversized," and mills pay about 30 percent less. This is painful! We like to grow big old trees and enjoy the forest as it changes through time. But in the last ten to fifteen years, most industrial timber companies have shortened the rotation of their tree harvests. Instead of harvesting at sixty or ninety years, most companies log at thirty-five to forty years of age. As a result, most sawmills have retooled and can handle only smaller logs. Shorter rotations meet quarterly profit goals but force small landowners like us to harvest earlier or face deductions on the scale sheet.

I discover that some oversized wood from Pacific Northwest forests is used for coffins in rural China. The invitation to the monthly meeting of the Lewis County Farm Forestry Association announces an upcoming talk by the owner of Millwood, a company in Olympia that ships logs for this purpose. I tell Dad we have to go. He gathers his thirty-year-old Society of American Foresters clipboard and puts on his city clothes. I check to make sure the T-shirt he wears under his button-down isn't frayed.

We drive to the Lewis County Courthouse and walk into the basement room where we have sat many times for talks about thinning, road building, planting alder, using herbicide, and surveying land boundaries. Dad is the "godfather" here; he and Steve both served as president of the association.

Rich Nelson from Millwood presents slides showing how wood gets from Tacoma to inland China. The wood he buys is big and rough and has many of the defects Weyerhaeuser will not take. Logs over thirty-three inches in diameter, spike knots, spangle—all these are acceptable for Millwood.

Provinces on the coast of China have outlawed burials; city dwellers must cremate their deceased. But in the countryside, families still want a coffin for a "soil funeral." Coffin design varies by region, but all require five thick boards about twenty-four inches wide. Many families prefer each board to be made from one tree; it helps the soul stay in the coffin.

I am pleased families want the otherwise rejected wood—material deemed unsuitable for plywood or two-by-fours—for the final resting place of their loved ones. The truth in myth trumps extreme rationality.

We have some hope that local markets for larger logs will reemerge. Recent carbon research shows that west-side Douglas fir forests absorb carbon at the highest rate between forty and seventy years of age. If timber companies receive compensation for lengthening their harvest rotation, and thus sequestering more carbon, small landowners like us could grow our trees longer too.

My family moved to this land in 1972 when I was nine. We'd lived in Seattle for the two years prior, and I invited my best friend from those days to the tree farm. We had just logged along the country road that bisects our farm, and I ducked my friend's head down in the car so she wouldn't see the stumps. Where did this instinct come from? How did I know at nine that city people looking at a cutover see only destruction?

Logging is a link in the chain that brings a natural, renewable resource out of the woods and into public, often urban space. It's almost as if this handoff, from rural people who tend and harvest to urban people who purchase and consume, is secret and silent. It ought to be a celebration, a merging of peoples, an acknowledgment and appreciation of gifts.

When I first moved back to the Northwest after years in New York City, I dated a man who couldn't understand why we cut our trees. His view was shaped by the timber wars of the 1980s and 1990s, a time marked by the overcutting of forests throughout the Pacific Northwest. I was excited to show him how a forest could be sustainably managed. But even after months of walks in stands of various ages, even after witnessing my family's demonstrative love for the land, this man thought we shouldn't cut another tree.

Language is part of the problem. We don't have a good name for our land. "Nonindustrial private forest" is too bulky and tells what we aren't more than what we are. "Timberland" conjures up lumberjacks or shoes worn by hipsters. "Tree farm" makes people think of Christmas trees. "The woods" is what Dad says when he's going out to mulch, tube, or thin, but it's not really a name. "Forest" conjures up a park—untouched nature. But when was "natural"?

Recently Dad and I hosted a student group from Evergreen State College. Before the tour, I asked them what came to mind when they heard the term "tree farm." Christmas trees, they said. Plantation, future paper, money, lumber, livelihood, monoculture. For "family forest," they said legacy, local community, family business, recreation, education, sustainability. But it sounds odd to say, "I'm heading to our family forest."

After the students leave, I walk in the house. On the door is a small metal plaque: Doug Stinson and Fae Marie Beck 1994 Western Tree Farmer of the Year. The American Tree Farm System gives this award yearly to families that demonstrate "the multiple use principles (water, wildlife, wood, and recreation) and their willingness to communicate the forest management message to many audiences." I realize that today's visit is just one of many that we've been involved in over the years.

Wood—harvested trees—is elemental. One of the five Chinese elements, wood has many characteristics including idealism, spontaneity, and curiosity. I used to have the sixth graders in my Parkrose classroom memorize Dorothy Parker's quip, "The cure for boredom is curiosity. There is no cure for curiosity."

One morning the rain ebbs and flows, slows and pounds on the metal roof above me; I lie in the bedroom of my mother's ceramics studio listening. When it's just a whisper, I can hear the hum of Peter's saw and the thump of his single-bit ax hitting wedges. I walk out to the logging site in the early afternoon with my notebook. On an old-growth stump sprouting huckleberry in its second life, I see Peter's coffee mug, which reads, "Sawdust is man glitter." The Canada geese call. The crowned kinglets chirp. When I am writing, I notice more, I ask more questions. Peter asks, "Are you writing a book?" Father and son both want to show me things.

I take a video of Peter falling a tree on the bluff: an old open-grown fir tree called a hooter because its branches are good for birds. We are leaving the hooter next to it. He is almost finished, but he waves me over and instructs me to climb up in the loader and knock the tree down.

It is cold and pissing rain. I use Peter's knee and a pile of branches to get onto the loader's tracks, then up the metal ladder. Young Peter tells me which joysticks move the shovel in the right direction. A few swings knock the tree to the ground. With the grapple, I move it to a clear place for easy delimbing. I am nervous, excited, out of my element. Totally new, these movements are large, not subtle. Expansive. Room for breath and thought and limit testing.

Peter also listens to the land. He writes poems in a series called "The Nature of Things" on his Facebook page. A poem he wrote about our land:

Shades of green and brown
on a moist mountain slope
inland from the ocean
far up a river
near the glacial till moraines
flat stone and pea gravel
stretch out far and flat

Brown thick bark of
Douglas Fir
stand deeply fissured black

The cracks of time
reaching upward to a pitch spout
running resin down
turned silver from the cold of winter
a frost that will not melt away.

Green moss
soft to the touch
is a blanket at the base
that rises up
to gray patches
of lichen cladonia macilenta
looking like clusters of
red throated trumpets
in the land of the small

I love the rhythm of the line "deeply fissured black" and the movement of the pitch in the third stanza. Peter lives with the trees as he falls them. It reminds me of ranchers who love the cattle they raise to slaughter.

The title of a book describing early logging, *They Tried to Cut It All*, gives a sense of the rampage and waste that comes to mind for many people when they think of the timber industry. Dad read this book when it came out in 1980; he saw firsthand the end of this era. In Grays Harbor County, the region whose history is the focus of the book, he worked in the 1980s buying timber for a mill he and his partner named "RHD (Root Hog or Die) Plywood and Lumber." He bid on Forest Service sales and organized the logging of thousands of acres, many of which were still lush with old growth.

One day after dinner, I ask him about that experience. Dad is sitting in his chair at the table; it has the best view in the house. Through windows on three sides you can see the bluff, the garden, the woodshed. About once a week he proclaims it a million-dollar view.

Dad tells me about a cut he once organized in the Olympic National Forest. The terrain was notoriously difficult, and he couldn't find anyone who would agree to build a road. Finally, a man nicknamed "Georgia Peach" said he could do it. The road was so steep each log truck required a bulldozer to push it a mile from the landing before it could run on its own. Mom joins the conversation; she remembers being on that landing. She stands up and imitates the logging boss, hand on her hip, strutting. "He had a gold chain peeking out of the shirt of his 'Aberdeen Tuxedo' and used binoculars to watch the men setting chokers 2,500 feet below. I told him, 'If the Lord intended this wood to be cut, the ground would be flat.'" Dad shakes his head and says, "I think I agree with you."

Around this same time, in the mid-1980s, Dad started questioning the federally set allowable cuts in western Washington. At dinner before a timber sale, he talked with a ranger in the Gifford Pinchot National Forest. "As the rain hit the windows and we ate our dinner, I said, 'You know, I get all over your district and it seems to me you can't sustain this hundred million a year.' The ranger replied, 'You're right, we can't. But the word is we have to "meet the cut." Other forest supervisors who have raised that question have been sidelined.'"

My brother used to say Seattle is Washington's biggest clear-cut. Portland is Oregon's. In working forests like ours, clear-cuts are not permanent. We will plant this year's cut next spring. By March, fir, cedar, and pine seedlings will be growing at nine-foot intervals across our twelve acres. We will start a new forest. Dad has been calculating the number of seedlings we'll need. A 1970s calculator, a yellow legal pad, and a sharp pencil are his tools. We've decided to plant western white pine, western red cedar, and Douglas fir. The cedar and pine are resistant to laminated root rot, present in our soil and exacerbated by climate change. Douglas fir is not, but we will plant it in the ashes of burn piles in hopes that some will survive.

2003. Two months removed from the cozy concrete and lively cynicism of New York City, I am at Camp Arrah Wanna on Oregon's Sandy River. Three other sixth grade teachers and I have brought our classes to Outdoor School for a week.

As the buses roll in, counselors and instructors line the driveway clapping and singing:

> "We welcome you to Outdoor School
> We're mighty glad you're here.
> We'll send the air reverberating with a mighty cheer."

Each child receives a wood-cookie necklace with his or her hand-written name and is assigned to an enviably older high school counselor named Maple or Weasel or River or Pebble.

While the students are getting to know their cabin mates, Coho, the camp director, gathers the teachers and field study instructors. We teachers are asked to come up with a camp name. I become Cowlitz. Coho then asks us to hold hands and compare our current feelings to a flavor of ice cream. My whole body rebels. I've been analyzing my feelings with a Freudian-trained psychiatrist on New York's Upper West Side. They can't be reduced to an ice cream flavor. And hold hands with strangers? I haven't made eye contact with strangers for ten years.

But I acquiesce and choose a meaningless chocolate chip mint. Much of the week is agony. Bunk beds, shared bathrooms. Loud dining hall meals. Campfire songs that drag me back to Baptist church camp emotions. No wine or beer.

But my students love it. They get to touch animal pelts, create a soil profile, examine nurse logs, put on plays, belt out songs like "There was a moose; he liked to drink a lot of juice," meet kids from other schools, and learn that the salt and pepper shakers are always passed together.

Each day, students participate in a different field study: water, soil, animals, or plants. My city students see how nature and humans interact. Most activities are well thought out, but one has a flaw that pokes at me.

Students build a model of a watershed with a city, a farm, a clear-cut, a forest, a road, and a river. They make buildings with sticks for the city and farm, "plant" ferns for the forest, and leave blank dirt for the clear-cut. Using a watering pail, they pour water on the landscape and consider which land use best helps protect clean water. They are given no information about impermeable parking lots in the cities or multiple applications of fertilizer and pesticides on the farms, and so without fail they choose the clear-cut as most responsible for polluted water. After the lesson, I point this out to the instructor and explain that a clear-cut is a stage of a working forest, not a static state. His eyes glaze over. On the final evaluation, I fill up the two lines I'm given and all the margins to try to clarify the problem. Two months later, I bring this same group of eleven-year-olds to the tree farm. The trees we planted in 1973 shade the road.

The trees grow, clean water flows, and deer, grouse, elk, and bear make the forest their home. The shade cools the salmon streams, and the growing branches act as carbon absorbers. No fertilizer, just one application of herbicide in eighty years of a stand's life. Dad's pickup has two bumper stickers: "Wood is Good" and "Family Forests are a Salmon's Best Friend."

In the months that made up the year after my brother died, I memorized poems. I repeated them aloud as I put one foot in front of the other on forest trails. Words written down and folded to fit into my fleece pocket. Rain-splattered, muddy, they live in the back of my notebook now—and in my breath and the back nooks of my mind—to be brought out when I need to ease restlessness.

A poem of Emily Dickinson's still resonates—I'm not sure whether it stays because of its darkness, or despite it, but there's a certain slant of light that oppresses like the heft of cathedral tunes. . . . When my mind is racing, my pillow overly warm, I can summon the rhythms of these words and breathe them down to my toes. Arriving at the last stanza:

> When it comes, the Landscape listens—
> Shadows—hold their breath—
> When it goes, 'tis like the Distance
> On the look of Death—

I've been brought deep into myself. I am in the shadows as they hold their breath, and I see the slant of light come and go, leaving its distant mark. This mark is on Steve's face as I lie beside him and Lou Jean the night he's died. I touch his cheek, tentatively. I lie down next to her as she snuggles him. She's put his wedding ring on a chain around her neck. His grimacing, groaning, and grasping—so hard to witness—have subsided to a look of softness, distant, but merciful. As the sun comes up, I walk back to the room I share with my husband, Tom. His warm body beckons from under the covers. I reach for him urgently. Tears wet our faces as our bodies merge. We must prove we are alive.

Emily's crisp rhythms allow me access into the room of death, and they create a structured entry and a way back out. The repeated soft "*ths*" soothe me and create a cocoon in which I can slumber. It's a fertile space, dark and warm and powerful.

On the day that marks a year, I drive with Lou Jean and Tom to each parcel of Cowlitz Ridge Tree Farm. We are in Lou Jean's black truck. From the rearview mirror hangs an eagle feather, a gift from a Hopi healer, one of the many healers on Steve's determined quest to rid his body of cancer. Next to my feet is a small shopping bag, "Taxco Sterling." Originally, it held "pretty shiny things" for Lou Jean from Steve's family forest advocacy trips to Washington, DC. Now it holds some of Steve's ashes.

We drive east on Highway 12. Limby fir and sprawling big-leaf maple line the road. Past the blueberry and tulip farms, up the hill to "Mossyrock," the largest of the timber parcels Steve loved. Lou Jean suggests we spread some ashes in the twelve-year-old alder stand planted in an old sloping field. The slope allows rain to slowly feed the alder roots. Sunlight passes through their oval leaves and patterns her tie-dyed T-shirt. Lou Jean reads a few lines of Wendell Berry, one of Steve's favorite authors:

And the world cannot be discovered by a journey of miles, no matter how long, but only by a spiritual journey, a journey of one inch, very arduous and humbling and joyful, by which we arrive at the ground at our own feet, and learn to be at home.

Elk have roughened the bark of the slender alder stems with their rubbing. I uncurl my fingers and spread my handful of ash over the previous autumn's crumbling leaves. My palm, now empty, is still covered in a gray gritty dust. What do I do with it? To wipe it on my jeans seems crude, disrespectful. I just keep my hand open and still, not wanting to disturb the dust's pattern along my lifeline.

Down the grassy road back to the truck, we walk through firs that Steve planted thirty years ago. It's easy walking; lower branches have fallen away, and the upper branches shade out underbrush. Only the sounds of ravens, hawks, and squirrels break the silence. As I lock the red gate painted last September with Dad, I pick a trailing blackberry. It stains my fingertips still coated in Steve's ashes. I lick them and make a fingerprint at the top of the first page of my spiral notebook. I am taking notes. I need this day to stay with me forever. I must own this day, to share with others who feel the hole of Steve's death.

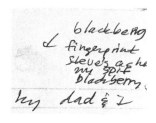

We head back west on Highway 12, windows down, hot air blowing our hair. Up Jackson Highway to "Callison," the last place Steve supervised a harvest of fir, hemlock, and cedar. Tom reads Berry's words: "arduous," "discover," "learn to be at home" float in the air with thistle fluff. I take another handful of ashes to the cedar log where Dad, Lou Jean, and I, earlier this year, ate smoked oysters and saltines, resting from our work of slash burning, boundary marking, and tubing newly planted cedar. I sprinkle my handful around the log so Steve can join us next time. Though I want them to stay visible, the ashes disappear into the dusty gravel and browned grass.

Callison is also the site of a yearly bounty of chanterelles hiding under waist-high salal. "Steve always found the most," says Lou Jean. We soon locate one of Steve's favorite mushroom-picking spots. I read Berry's words: "journey of one inch," "humbling," "joyful," and they mix with the ashes. Here, the dust from my ashy palm stays visible on the green pointy leaves of Oregon grape and salal. After Lou Jean spreads her handful she shows me, in her open palm, a small screw. She says, "It's a part of Steve. From his broken ankle. I found the plate earlier."

Steve's death is a raw wound in wet ground after a hundred-year-old cedar falls in a windstorm: broken roots, exposed rocks, soil that has not seen the sun for decades. But a year has passed, and in the hole from the wind-thrown tree, there is new growth—stronger bonds between those gathering round. The new growth is not a consolation. It just is. And this growth is nothing like the tree, but without the destruction, it wouldn't be there. We will tend the tender shoots.

Simple berry buckle: butter, eggs, sugar, lemon, flour, and berries. Picking the berries is the best part—it brings Steve to me. He died with blackberry scratches on his hands. And the vines still grow, trailing blackberries low to the ground. New vines stretch along the edge of a clear-cut now filling with four-year-old pine, fir, and cedar along with foxglove, mullein, and daisies. It was Steve's favorite spot and is still full of fruit; next year the shade from the growing trees might not let in enough sunlight.

My favorite trove this summer is a broken-down cedar log, decaying for fifty to sixty years, its red flesh webbed with vines, each bearing five or six plump berries. A pileated woodpecker drums nearby; a hummingbird whirs in the fireweed. I'm so determined to pick a berry hiding under a salal leaf, I get nettles on my chin. Steve feels near and I can almost hear him chuckling.

An hour of picking and my container is full. I walk back home in my hickory shirt and Carhartts, my arms and legs unpierced by the thorns. The berry juices paint the batter red. The oven that baked Steve's last pie plumps and goldens the cake. Our fractured but mending family digs through cream to the warm buckle and finds Steve's berries from the young forest.

2014. After the hospice nurse had come and gone for the day, leaving us with new medication, Steve wanted to go pick berries. We were hopeful that the medication would reduce the fog and confusion his diseased liver had been sending his brain. My sister Julie, her son Griffin, and I drove Steve in the farm's Ford Tundra out to his favorite berry-picking road. We stopped among dense, bushy twelve-year-old firs where berries crawled up spindly trunks, around the competing salmonberry, and rode on honeysuckle vines.

Steve pushed into the branches and underbrush, his green flannel shirt and black sweat pants just visible. We picked alongside him, worried he might fall. He didn't talk, just reached out his beautifully sculpted hand again and again to get the next berry. If he felt the thorns scratching his jaundiced skin, he didn't say. His bucket full, he asked to go to another spot where Mount Rainier stands visible on a clear day. Here were few berries; mostly wildflowers and grasses. We collected a few for Mom to identify. I picked a grass with a purple stem, and Julie found a white flower with six petals. Steve picked up a Vexar tube that had been used to protect a young cedar and said, "Mom will know the name of this one." My body and mind moved heavily as we put the tube in the truck.

We'd picked two quarts, enough for a pie. Back at the house, Steve rested as I stirred sugar into the berries and gathered the ingredients for his piecrust recipe. When I called him in, he looked at me with uncertainty in his eyes. He added the ice water straight into the flour, the cold butter still next to the bowl. My stomach dropped. Why had I thought he could do this? I felt like I'd asked a struggling reader to read aloud in class. As I stood, chilled with despair, Lou Jean walked by and I waved her in. She took one look at the counter, at me, at Steve, and mixed the wet flour, added some magic with the butter, and I was able to roll out the dough.

Steve ate two slices smothered in whipped cream, a huge smile under his chemo bandanna, his scratched hands holding the fork.

It's been one month since Steve died. Dad, Lou Jean, and I spend our days in the woods with Dad's farm dog, Tal, and Steve and Lou Jean's just-adopted Bouvier, Izzy. Tal is only six years old, but something is wrong. Her usual smooth leap onto the tailgate ends in a dreadful crash. She howls in pain. Dad bends over, pats her head until she is calm, and helps her into the back of the truck. We drive to Mossy-rock to take inventory of a twenty-year-old stand that got away—wild cherry and big-leaf maple instead of Douglas fir.

On the way back, Lou Jean and I decide we'll take Tal to the vet. Her mouth is swollen and she has a lump under one ear. The vet looks at Tal, looks at us, and shakes her head. "It's an oral melanoma. Not treatable."

We've failed to notice Tal's growing weakness or her once-black coat tinging brown. Too much death. We bring her back home and tell Dad. His shoulders sink and tears trickle from his deep-set old eyes.

For the next few weeks, he and Tal are inseparable. On trips to town, to Callison, to Mossyrock, to Finn Hill, or even just down the driveway, Tal is by his side. When she drags into the yard a mess of stinking deer parts some hunters left, he doesn't scold her.

Mom and Dad go to a family wedding; Tom and I stay at the tree farm to take care of Tal. Tom buys a whole package of hot dogs and feeds her one each morning and evening. Tal seems to know this treat is significant and nuzzles Tom with big sad eyes.

It is time. Dad has chosen the spot: the Douglas fir nurse log where Steve, while leading tours, would pause to tell visitors about the life cycle of the forest. Dad digs a hole in the dirt, home to insects, mushrooms, and roots, then brings Tal and a gun. A gun in the living forest. Next to the hole, Dad sits and pets Tal, looking her in the eye and thanking her for the years. He pulls the trigger and there is blood. The bullet hits Tal's head and her body sinks into the hole. Dad fills the hole with dirt and places a cedar post in the center. He's sawn the post from a hundred-year-old wind-thrown bee tree and now decorates with nails spelling TAL.

I am in awe of my father's power, his capacity to be raw. The intimacy and the violence mix my emotions. It is out of sync with urban twenty-first-century sensibility. I am wary of telling the story.

Low gray clouds meet the treetops edging the driveway and hover above the log cabin. Lou Jean has lived in the cabin since Steve's passing. Just a few hundred feet from Mom and Dad's, she can easily visit or share a meal.

We've been working together to make the tree farm ours, walking the boundary lines and learning ways to help the trees grow. Dad is our guide. He loves having new people to teach. I feel fresh energy about the tree farm; curiosity and knowledge invigorate me.

I walk up the driveway. Lou Jean is on her couch, a crocheted afghan wrapped around her legs. Her dog, Izzy, lies near the wood stove, eyes steady on his owner.

I've brought annual report documents and start talking, but Lou Jean interrupts me. "There's something I have to tell you." I look at her, my stomach constricting.

CT Scan
Dark spot

Blood tests
Cancer
Colon Cancer
Chemo
Radiation
Surgery
Chemo
95% survival rate
Outrage fills me.
It's supposed to be a year
of healing for Lou Jean.
Too much.
Overwhelming.
Pouring salt on a wound.
Adding insult to injury.
Beyond the pale.

We cry, hug, and walk down the driveway to tell Mom and Dad.

I'm out with Dad in five-year-old cedar. We've tubed each tree to protect it from deer browse. It is a gloriously sunny winter day. Lou Jean
is in Philadelphia seeking the least invasive cancer treatment. I take
out my phone and record Dad as he talks to her. He's dressed in his
"Aberdeen Tuxedo"—the usual hickory shirt, jeans, and suspenders.
Today his hat reads, "Life Members Smokejumpers."

"Hi, Lou Jean, we're sending this to you with love and white light.
We're down in our lovely cedar along the creek at Finn Hill, down
below the twenty-four-inch culvert. These cedar are really doing great.
We're cutting the tubes on one side so they can expand their wings and
fly. We've got some that are eight feet tall now. It's really too bad you
aren't here sharing it with us, but we send it to you from the beautiful
state of Washington. We love you, girl."

The doctors in Philadelphia successfully treat Lou Jean and after
a year's recovery she is back out in the woods with us. I write a poem
in celebration.

Across Collins Road
from the driveway and paper box
is the well-traveled trail
to the spring
to the 5 year pine and cedar plantation
to the island of 70 year cedar and fir
to the western boundary
to the wetland with Ash and Camas

Lou Jean and I are the new owners
of this family land
bought in 1971 by my dad,
our current senior forester.
As Raven, Eagle and Red Tailed Hawk,
Steve watches us
Lou Jean's husband, my brother

This year we find two treasures on the trail
as if the land wants to
welcome its new nurturers
First, an edible looking leaf
abundant in the boggy swamp

Parsley-like leaves
We taste it.
No numb lips or tongues,
Our phones tell us it is lovage.
We gather more.
In the cold mud our rubber boots sink to mid-shin.
Cottonwood buds are beginning to burst above
I make lovage pesto
with olive oil, walnuts and lemon
Lively, bright green sauce on friend-caught halibut.
A few weeks later, on the same trail
we spot a brown ridged cone
pushing up through last autumn's leaves
A mushroom, but what kind?
Chanterelles don't grow on this trail.

Lou Jean thinks morel, and after two
YouTube videos contrasting true and false morels
We race home to gather knives and a basket
giddy with our find.
Our land grew it!
We found it!
Our knives slice the pitted, plump caps
from the hollow stems.
Fifteen crowns fill
one of mom's ceramic plates.
We take photos before slicing
to share with those not present at tonight's table.
Butter, garlic
heat, cast iron
spatula
toasted country blonde
imported from Portland
We bite and hear a red tailed hawk
sing with pleasure.

It's 2002, my eighth year in New York City. I've completed a master's degree in East Asian studies, gotten divorced, worked an office job long enough to get restless, and decided to try public school teaching. In the spring I take my sixth grade students across the Metro-North rail line to the New York Botanical Garden. We are going to read poems under a cherry blossom tree.

We have written list poems based on the writings of Sei Shonagon, a tenth-century lady-in-waiting at the court of the sixty-sixth emperor of Japan. She wrote *The Pillow Book*, a collection of poems, lists, observations, and court gossip. I'm hoping her snippy observations will amuse my students, who spend much of their time in friendly and not-so-friendly insult exchanges.

My instructions at the top of the handouts: "Read the lists. Choose an item that interests you. Copy it. Illustrate it. Add an idea of your own that fits the concept of the list. Illustrate it."

Sei Shonagon's List of "Depressing Things":
> A dog howling in the daytime.
> A cold, empty fireplace.
> A letter that has been returned.

The family of someone who did not receive a hoped-for job.
A badly painted fan.

Sei Shonagon's List of "Hateful Things":

You've just settled sleepily into bed when a mosquito announces itself with that thin little wail and starts flying around your face. It's horrible how you can feel the soft wind of its tiny wings.

A baby who cries when you're trying to hear something.

Someone who butts in when you're talking and smugly provides the ending herself. Indeed anyone who butts in, be they child or adult, is most infuriating.

And I hate people who don't close a door that they've opened to go in or out.

Sei Shonagon's List of "Things that make your heart beat faster":

A sparrow with nestlings.

A fine-looking gentleman pulls up his carriage and sends in some request.

To wash your hair, apply your makeup, and put on clothes that are well scented with incense. Even if you're somewhere where no one special will see you, you still feel a heady sense of pleasure inside.

On a night when you're waiting for someone to come, there's a sudden gust of rain and something rattles in the wind, making your heart suddenly beat faster.

Sei Shonagon's List of "Splendid Things":

Chinese brocade.

Ornamental swords.

Long, richly colored clusters of wisteria blossoms hanging from a pine tree.

Grape-colored, figured silk. Violet is a splendid color wherever it is found—in flowers, in fabric, or in paper.

Snow lying thick in a garden.

In the gardens, we read our poems. Cherry blossom petals, pink against the gray sky, fall onto black hoodies and white sneakers. We eat our lunches, Tupperware containers of chicken and rice, Lunch-ables, and Gatorade, and climb the rocks, happy to be free of desks and four walls.

As we walk back to the school, twenty-five twelve-year-olds in a scraggly line, I notice a middle-aged man glower and move out of our way. I comment that he must not like kids. One of my students, Luis, runs up to him and shouts, "I'm a kid!"

It is a successful day. Many of my teaching days in the Bronx have not been. Before walking into my first classroom, I hadn't talked with a twelve-year-old in years. The students think I'm ancient and can't believe that at thirty-eight I don't have any children of my own. The New York City Teaching Fellows' eight-week training program has left me woefully unprepared for a middle school in the Bronx. I am a soft soul in a hard place.

One day, a girl is angry with me; I've made a new seating chart and she is no longer next to her best friend. She yells, "You're horrible. It's no wonder you never found anyone who wanted to have kids with you." I gasp. My hands shake. I am scared. A scared teacher is blood in the water. I had been a "good kid," a Goody Two-shoes to some of my peers. I didn't and don't understand or ever expect meanness. The sharks circle.

But I love the intensity of the classroom; the challenge of engaging disparate learners, finding just the right book for each reader, teaching stories from Homer's *Odyssey* and having a student, one who was often on the floor under his desk, pop up and say, "Oh! so Zeus was a player!"

At Christmas they convince me to have a party. I'd envisioned food and games, but the day of the party, the girls push the desks to the walls, turn up the music, and start dancing. The boys sit at the tables playing dominoes. After dancing to Jennifer Lopez and Destiny's Child, the girls put on Luis Vargas and invite me to their dance floor to teach me the bachata. My feet master the simple steps, but my hips have no rhythm. A student giggles, "Ms. Stinson, your mama didn't dance with you when you were a baby!"

In the winter, Luis draws a picture of his neighborhood block.

I still have this drawing. A snowman is dressed in a polo shirt. Luis's friend Julio wears a NY Yankees cap and a Sean John shirt. His voice bubble says, "Blood For Life." Luis wears a red bandanna tied around his head and his bubble says, "DDP (Dominicans Don't Play) for Life." Above all the figures rise brick buildings with many small windows. Luis titles the drawing "Ghetto Snowman."

I keep the drawing in an old photo album, next to a photo of me with students hanging over my shoulders and tugging on my arms. My body is slumped and a tired smile tries to brighten my face. A caption reads, "Don't Let This Happen To You." The yearbook teacher who took the photo and added the warning always had her students in quiet lines.

I begin to realize teaching in the city is not a sustainable profession. And my social circle has shrunk; friends have moved and my search for a man has dead-ended. My family in the forest is calling me. I've made friends with many of the trees in Central Park and in Fort Tryon Park, but they are not holding me as close as I need to be held.

I sell my apartment, buy a VW Passat, and drive the three thousand miles west. The last day on the road, I follow Highway 12 from Missoula, Montana, to Steve and Lou Jean's driveway. Lou Jean takes a photo I still have hanging on my wall: my torso bursts out of the car's sunroof, my arms reach to the sky, and a smile takes over my face. I don't have a job or a house, but I am home.

One month later I am teaching again, now at Parkrose Middle School in Portland. Here there is a basic level of respect between the administration, teachers, and students that was lacking in New York, and this makes my new classroom a place I can thrive. And while I loved the mostly Dominican student body in the Bronx, Parkrose is delightfully diverse: students from Vietnam, Mexico, Chuuk, Russia, Samoa, Somalia, Eritrea, and other countries join African American students from gentrifying North Portland and White families with deep local roots.

I want my students to feel known. On the first day of the year, I put an "All About You" questionnaire on each student's desk. Anxious teenagers have something to do as soon as they walk in, and I gain a treasure trove of information. That night, I use their answers to make notes next to the eighty photos on my three seating charts: "likes to draw," "plays soccer," "has three younger sisters," "just moved to Parkrose," "home language is Hmong," "wants to be called C.J.," "likes to fix old motorbikes," "good at baking." Over the next two days I plan enough independent work time to go to each student, kneel at his or her desk, and ask a question or two.

I find a teacher next door with whom to excitedly share lesson plan ideas. During our thirty-minute lunch we fine-tune the steps of an activity. In the evenings, we text each other new resources we think will cultivate our students' curiosity. We notice which texts help students find themselves and connect to a larger world. I become a leader of the Professional Learning Communities where teachers gather and collaborate. It's good hard work.

But the students' desires and needs push against classroom walls. I love them and am overwhelmed by them. So many bodies in one room: the aroma from my thermos of Earl Grey tea mixes with AXE body spray and shower uncertainty. I ask myself whether I "own" my room enough. Budget cuts, high-stakes tests, and growing poverty all add stress to an already stretched-thin institution.

I wonder why we educate in a cube. One morning before the students come in, I look around the room and take note of the four walls. On the front wall, a whiteboard with the day's learning objectives is screwed on top of the old chalkboard. To the right a yellow-laminated Lock-In Poster tells us to get low to the ground, away from windows, and stay silent when we hear two long beeps.

On the back wall, I've made a timeline with photos: Anne Hutchison leading a Puritan meeting; John Ross, chief of the Cherokee Nation; Lincoln at Gettysburg; Robert Smalls on his stolen Confederate ship. Next to it is a list of students' questions about Sherman Alexie's *Absolutely True Diary of a Part-Time Indian*: Why does Junior draw cartoons? How is Reardon High School like our school? What do you think about Junior's dad? Other questions and comments from yesterday's lesson on the Mexican-American War flow through my head: "How do we know John Riley switched sides in the war?" "It was so long ago, why should I care?" "Library of Congress?" "They keep papers there?" "My job will be to make a fireproof building to protect them."

The left wall is a bank of low windows: grass and daffodils, Mount Saint Helens in the distance, room for daydreaming. In the left corner hangs the American flag. I roll it up each summer—it's on the June checklist under "hand in your keys."

Next to the door in the right wall sits the breakfast cart with 30-percent-less-sugar Cocoa Puffs, cellophane-wrapped waffles, and flickable raisins. A student petition to "Get Our Muffins Back" will soon adorn the cart. A bookshelf filled with graphic novels, fantasy, adventure, sports, horror—books to match all kinds of readers—stands in the corner.

During my eleventh year at Parkrose, Steve's illness is ravaging his body and I decide to resign to be with him.

On the west edge of the new stumps lies the "Snag Patch," nine acres of
fifteen-year-old fir and pine. Green leaders and branches grow free
above a thicket of Indian plum, hazelnut, and snowberry. At eye level,
entering the stand is daunting, but I'm going in to look for honeysuckle, an unsuspected enemy of the fir.

I am by myself. I know what to do. I don't need Dad to explain.
I relish the solitude, the independence, and I work at my own pace.

My tool belt holds black-handled Stihl clippers and a Zubat "silky"
pruning saw. My gloves are a wonderful size small. I am smaller than
anyone in my family and for years wore gloves too big for me. There
is a pleasurable confidence in wearing gloves that fit.

Fibrous honeysuckle vines crawl clockwise around the slender firs,
strangling their trunks. Individual vines, like upside-down ribbons on

a maypole, find each other, twine into a ferocious braid, bulge and deform the bark. Parasitic roots dig into the bole and suck the hard-won nutrients from the adolescent tree.

I look up each tree for a vine, just now budding green leaves. They will be easier to spot when their beautiful but ominously scentless orange blossoms bloom.

I clip and pull, feeling powerful and wishing it were as straightforward to release a human adolescent from the strangleholds of poverty and racism.

> Trauma from ten homes in five years?
> Clip, snip
> You can start to breathe.
> Bruised from the blindness
> of teachers?
> Untangle, unsnarl
> stand straight and tall
> and vigorous.

Moving from one tree to the next is work. I push my thighs and hips against the underbrush. Himalayan blackberries tangle my shins and tear the hat off my head. I climb over branches and decomposing logs. Thorns rip the skin through my jeans. Tomorrow, purple puncture marks will dot my legs. Dead leaves, needles, and twigs sift down my shirt and into my bra.

I take a break, shake out my bra, and sit on a stump from 2006. Next to me is a snag Steve created during that harvest. Woodpeckers and insects feast on its decomposition. At an angle in front of me runs an animal trail. I wonder when the last animal used it. How many hours ago? Which animal? What was its purpose? In the summer, deer have to go all the way to the Cowlitz River for a drink. In the winter our wetlands provide enough.

Young nettles poke up through a layer of reddish-brown pine needles. A last year's bird nest snugs against a young trunk. The wind blows high in the trees. Nearer the ground, just the needles on the fir move. Not the trunk, not the branches; just the needles dance, and the Oregon grape leaves twirl next to my knees.

A whoosh from nearby Gemini Grove meets with a soft rush from the ridge. A frond of last year's bracken fern, dried and folded over a lower tree branch, starts waving. Brown papery pieces float down and land on my phone as I type.

I wonder how it is that I am here, alone with the trees and not in a classroom swirling with fourteen-year-old energy. It's a relief to my psyche not to be responsible for ninety different human beings each day. Sometimes I get restless, thinking I need more human stimulation, but for now the forest is a place for me to grow myself.

Until Steve's illness, the tree farm was my weekend place, a place for family dinners and holidays, the trees a backdrop, not the stage of my life. I would walk the trails but not explore or think about how to keep the forest healthy. Now as I work, I listen to the woods, hearing stories, and I want to tell the world what I hear.

Done resting, I move to release another tree. On its lowest branch is a small owl. It just stares at me with eyes like amber disks. I take a video. I walk away. I come back and still it stares. A brown fluff of feathers covers its six-inch body; a few white ones pattern its chest.

Many twirls of honeysuckle vines choke the bole of the owl's tree, but I cannot disturb it. Back at the house, watching the video, I hear myself whisper to the owl, "Are you real?"

The white eyebrow V, lack of ears, and small size identify it as a northern saw-whet owl. Wikipedia tells me the female saw-whet lays five or six white eggs twice in one season. She uses abandoned woodpecker holes and tends her offspring until fledged. She then leaves them to the father's care and goes off to find another mate. This is called "sequential polyandry." I learn that early loggers gave this owl its name because its sound reminded them of the rhythmic tone made when sharpening a crosscut saw. The Lower Cowlitz word for owl is *chi-nuk-shitm*. I want to know what their name for this owl is.

Saw-whet owls choose dense young timber for their home. I feel proud that we have created their preferred thicket. And this thicket is just a few years away from entering the most robust carbon-absorbing stage of a forest's life. During the next several decades the trees in the Snag Patch will absorb more carbon per acre than those in our beloved Gemini Grove.

Over the next few days, the owl's eyes come back to me, amber disks, piercing. I walk a path near the owl's tree and imagine her nesting in a woodpecker hole, tending her young, and choosing another mate. I have new knowledge of life in what seems a tangle of undergrowth and struggling fir. The owl saw me and accepted my presence in its home. I want to meet more of the forest's creatures.

I'm up on the ridge above the Snag Patch, a trail running ahead of me, the wet grass flattened by deer, elk, porcupine, and skunk. On either side of the trail, young pine, six to ten feet tall, have been stripped of their protective bark. Vulnerable fleshy cambium exposed, sap drips in the open air. I need to know more about porcupines and begin researching.

A female porcupine gestates for 202 days and gives birth to a single young.

On one tree, month-old scars have already blackened. Splotchy wounds wind around the bole. Newly gnawed stems, on the next pine, still shine white; teeth marks give evidence of the nocturnal perpetrator. The peeled outer bark lies scattered on the ground; porcupines like only the softer inner layers. I begrudge them this fact. If they're going to kill our trees, they shouldn't be wasteful.

Like all rodent's teeth, a porcupine's two large front incisors will continue to grow throughout its entire life.

Just next to a healthy cedar, a porcupine has climbed midway up a tree, sat on one whorl, and stripped the bark all the way up to the next green whorl, ten inches, the height of one year's growth. For each of these trees, there will be no more growth. The green of this tree and all the others on this ridge will be red next summer and black the next.

In winter the inner bark of trees forms the main part of porcupines' diet. They chew through the outer bark of trees to get to the edible cambium.

Home from my walk, I tell Mom about the porcupine damage. "I used to trip Dad's traps, hating the thought of dead animals in the metal claw. But now, I hate those porkies. Seeing those girdled trees— makes me cry."

"Those fackers," Mom replies.

The moon is bright as I drive home from a Family Forestry Foundation meeting. I find Mom and Dad in their respective chairs next to the wood stove. I put my bag on the dining room table next to notes I've taken about porcupines. While I want them gone from our forest, I want to know more about them.

I warm myself by the wood stove and look at Dad. "It's a perfect night—let's go see if we can get a porcupine." He is reluctant, having settled in for the evening, but Mom says, "I'm interested in this nighttime effort. I've always endorsed hunting with a spotlight."

Dad and I look at her in surprise. Is she serious? Mom doesn't usually like to have anything to do with hunting. But then I realize she is doing it for me—she feels my anguish about the damaged trees.

And she's made Dad smile. "That'd be great, Fae Marie. You know, it's the same shotgun I used to shoot the goose that brought us together in 1958."

Mom fills in the missing links. "Our friends in Ketchikan, Carl and Joan, arranged a goose dinner and invited Doug and me. This was after they introduced us at the Methodist church. Doug worked with Carl for the Forest Service, and I taught with Joan."

Porcupines can live up to eighteen years. They do not build a nest but sleep in a tree or find logs or rocks to sleep under. They have a home range of about twenty-five to thirty-five acres.

Dad gets ammo from a red-and-yellow cardboard box: Super X Mark 5 Plastic Shotgun Shells. Mom puts on her insulated boots and adds a denim shirt jacket over her house layers. On her shoulder is a safety pin, there since Trump's election. I put on my blue rubber boots and lift myself over the tailgate into the bed of the pickup.

Dad climbs up and stands next to me with his loaded shotgun. Mom pulls herself into the driver's seat. We lean against the toolbox as Mom starts to drive. She's unused to the pickup and it lurches, throwing Dad off balance. "God dammit," he yells as he braces his legs. I see Mom wince.

Dad has plugged a spotlight into the cigarette lighter and threaded the cord through the open back window, and I throw the beam onto the sides of the road as we move slowly through the woods to the ridge. I touch the wooden stock of his shotgun, and Dad says, "I bought this at the commissary in Yokosuka, Japan, in 1957. I knew it was a good duck gun, a twelve-gauge pump—just one shell at a time—and I guess knew I was going to live somewhere with geese and ducks."

The front of a porcupine's body is covered with long, yellowish guard hairs, while the back and tail are covered with up to thirty thousand quills that are interspersed among dark, coarse guard hairs.

The moonlit night suits our purpose; wet quills make bad defense, so rainy nights keep porcupines in their dens. Orion and the Big Dip-per grace the moonlit sky. The cold wind numbs my face. Mom sits up tall to see over the steering wheel; her once five-foot-eight frame has shrunk to five foot three. She avoids the potholes and stays on the road.

We stop just past a stand of older white pine and cedar, where we've seen the most damage. Mom stays in the truck and follows the Blazer game on her phone. Dad and I climb out, I with a huge yellow flashlight, and he with his sixty-year-old shotgun.

Porcupines are nearsighted and slow moving. They are solitary, denning with other porcupines only during severe weather.

Walking from the road to the ridge, we look for movement on the ground or up in trees. The flashlight beam catches the afternoon's raindrops on the tips of pine needles and makes the scarred trunks glow. We move to another area to check our traps, our footing unsure on the slick grass. Ignored apples, peanut butter, and rock salt lie next to untripped metal traps. Dad mutters, "What do they want, caviar?" We see no rodents. No glinty eyes, no quilled humpback waddle.

No shots echo into the night air. More sharp teeth will tear the trees' flesh under this moon.

This spring's lovage is wildly abundant, growing thick and green in slowly flowing water. The whole forest is burgeoning: miner's lettuce, fiddle-headed sword ferns, red alder catkins, soft nettle leaves, ponderosa pine "candles," silky green tips of cedar.

This burgeoning plucks a chord of sorrow deep in my soul. I do not have children. I do not have a child. My body has never swollen with new life. Working in fecund woods where everything reproduces, all the time, without thinking, brings my empty womb and quiet house into painful convergence.

Often I wonder how this happened. How is it that I don't have a family of my own? The saw-whet owl mates twice in one season! I didn't manage to mate in thirty seasons.

The timing of my life choices didn't correspond with the seasons of my body. At twenty-seven, I married a man who was never quite ready to have a child. I divorced at thirty-five and didn't remarry until forty-four.

But still, it took a lot to keep nature from running its course. Why did I cultivate religious and societal pressures that kept me from reproducing, from creating new life?

Where were my clippers when I felt the strangle of my own honeysuckle?

When I was thirty-seven, I found some courage and sent this letter to a man I had dated:

Dear Bob,

I've been thinking about your offer of "getting me pregnant." It has been simmering in the back of my mind amid this week's visit to the doctor, my sister's move to the Stinson-home base, bad dates, and good conversations with the smart, nonlinear, jock. Before I consider it much more seriously—is the offer still good?

I want a child. I would love to have your child. I would like to raise him/her with my brother and sister and their spouses and my nephews. And of course the eccentric grandparents. I would like him/her to be able to visit you now and then—to know who you are—what are your thoughts about what kind of arrangement would work?

Love,
Ann

Bob called and wondered aloud what complications might arise if the child (a daughter he said it must be—he had three sons) didn't want to live on the tree farm. But he invited me to Florida for Thanksgiving. I accepted. The beach was spectacular, but sex was awkward. Somehow we didn't work. At brunch we ended up at a table next to a family with a beautiful baby girl. I burst into tears that wouldn't stop. We left, but in my distress I forgot my sweater. I chose a soft chair on the condo balcony and sobbed while Bob went back to get it. When he returned, he had a large Dunkin' Donuts coffee—light with no sugar, just how I like it. The coffee stemmed my tears.

Decades later, Bob and I are still friends; he tells me about his grown sons and I tell him about the farm.

I immerse myself in the new forest life, glad to have something to tend. Ferns and cedar boughs buffer the hole in my heart, pad the gully, and soften the ache.

This spring, I am also tending Mom. Today she sits on my deck in Portland. Her pajama top overlaps a scarf tied around her waist acting as a skirt. It hides the catheter and bag of urine. Two days ago she had bladder prolapse surgery, and we've been at my house instead of on the farm. Mom moans, groans, and grimaces. I deliver pads, MiraLAX, and pillows. Finally some pain meds.

She turns to me and says, "It breaks my heart and I know yours too that you didn't have children. You are taking care of me. I can feel that strongly. Thank you."

From the mudroom I hear Mom playing her spinet. My boots tugged off,
I stand behind her. Handwritten notes on the music, Bach's Prelude
VIII, read:

> Prelude: Nov '77, August '89, July '98
>
> Postlude: July '95, January '99, June '04

For thirty years, Mom was an organist in local churches, first at
St. Stephen's Episcopal in Longview and then at Westminster Pres-
byterian in Chehalis. For all those years, Mom had a pipe organ in
our home in the forest. When I think of that now, I'm amazed. My
parents were frugal—we drank powdered milk, Mom made bread and
our clothes, the meat on the table was venison we cut and wrapped
on our Formica counters. But Dad likes to please Mom. When we

first moved to the land above the Cowlitz River, he logged three-fifths of the standing timber, paid for the land, and with the first profit bought her the German-built pipe organ. Grenadill, a dark African hardwood, and ivory made the two manuals. Six different ranks of pipes stood straight behind carved oak shades. Mom used the organ to practice for church services and to wake us up in the mornings with a rendition of "Irish Washerwoman." She loved congregational singing, and to accompany a group of people as they lifted their voices in song gave her great satisfaction. The spinet, a smaller version of a harpsichord, is the only instrument Mom has now. She and I still sing hymns together as she plays—her aging hands can still reach the scope of an octave, an action difficult on a piano or organ. This morning, after Bach, we sang "There Is a Balm in Gilead" and then listened to Nina Simone sing the spiritual. We tapped our feet and moved our hips to the syncopated beat.

In 1995, still an organist, Mom started a new creative endeavor—she became a potter. After a goddess pilgrimage to the island of Crete, she began making ceramic objects. She was sixty-one.

She makes her own goddesses, foot-high clay women with shawls and skirts, hips shifted away from the waist, back straight and tall, no eyes. Sometimes with hats, they are always mystical and pensive, brave and powerful. She makes masks and hollow rocks and totem poles. She makes vessels. Vessels to hold water or wine or cigarette butts or flowers and pussy willows.

In the months after Steve's death, Mom made a vessel she calls a pietà. The pietà, in Mom's case, has two eyes, but otherwise no human features. No obvious mother and dead body. It is a vessel holding a vessel; a round body with a long, thin arm holding a smaller open cup. A long, slender neck rises high above the round body.

On a May afternoon, we sit in the carport at the farm, next to the blue door leading to the freezer, the sauna, the canned goods, and the half racks of beer: Blue Moon for Mom and BridgePort IPA for Dad. One of the first sunny days this year, it is glorious. Dandelions and heather and lilacs and tulips are in full force and rosebuds already signal summer.

Mom sits in her customary chair, on her blue cushion. It's the chair she sits in to smoke her Marlboro Lights, to follow the Blazers' games, play Words with Friends on her iPhone, monitor the progression of the seasons, and watch the birds. Next to her chair on a wooden box sit two glass jars. One holds a lighter, a notepad, and a pen for an occasional haiku. Wads of silver paper—cigarette pack linings—fill the other jar, awaiting an art project.

I sit on the concrete, my feet in the gravel. She "tsks" and says, "A man once told me sitting on concrete is bad for your parts." I sit anyway.

Mom tells me the pietà was inspired by the turtle goddess of Myrtos. "The Greek figure with the pitcher in the arm seems to me a parallel to the central concept of a traditional pietà, the mother holding the dying Jesus. I mean, this figure is cradling a pitcher in her arm. In my thinking, it symbolizes the dead child. Religion, such as Christianity, provides a vocabulary, an iconography to express things that are difficult to express."

As she sees it, the mother and dead son are "communicating without words. In the original Greek figure, the water, liquid can pass from the mother shape to the child shape, the pitcher. I closed that passageway to mark the ending of the relationship. I now know the relationship with Steve doesn't end. I could have left the hole open. I didn't realize that until quite a while after I finished the piece."

Mom often says that when she works with clay, her mind is free. And working with clay has enhanced her relationship with music. She realized her music had just been note by note. "It had no heart. I was just interested in making sure I played the right note with the right rhythm. After clay, a spiritual quality began to infuse my musical performances."

I'd thought a pietà referred to the Virgin Mary and the infant Jesus, not the body of Jesus. I tell Mom she's neglected my education and wonder when she learned the term "pietà." She didn't learn it until 2000 when she and her organ teacher were in Italy visiting churches and they saw Michelangelo's *Pietà* at Saint Peter's in Rome. Later, on a visit to New Harmony, Indiana, she saw a piece entitled *Pietà* commissioned by a woman who had lost a child. Stephen De Staebler created an abstract metal sculpture of a woman with a head coming out of her chest. This sculpture inspired her to think of her own interpretation of a pietà.

For Mom, the image of the pietà connects her to the power of women. It makes her think of an Equal Rights Amendment rally where she heard Sonja Johnson invoke the power of our ancestors starting with Sojourner Truth. Mom's 1930s and 1940s upbringing in the panhandle of Texas left her hungry for stories of powerful women. She wasn't alone. When she was on the goddess pilgrimage, she tells me, one of the women said, "We Protestants didn't even have Mary."

We walk to her clay studio just across the driveway to see the image she used as a model for her pietà. In her tie-dyed socks and garden shoes, she crosses the room to a file drawer I've never noticed. Old file folders lie piled in the drawer, reused but not relabeled. The old labels read: "Current Correspondence," "Enron," "Art Clippings," and "Modigliani." All contain art clippings. I smile as I realize that Mom, whose house files are tidily organized, allows herself disorder here.

From a file near the top, Mom pulls out a goddess image of Myrtos. It is a paper square, cut from a computer printout advertising the goddess pilgrimage. Mom's printer tray is often full of once-used sheets, and she's printed the turtle goddess on the back of a logging receipt from March 6, 2014. It records two loads of logs from Baldwin Logging totaling 9,430 board feet.

I ask what drew her back to this image when Steve was dying. She says she'd made a few before based on some by Picasso and then sighs, "I don't know. The answers to your questions are visceral. They skip my brain. I can't articulate it. How about 'The Lord led me?'" I laugh out loud. I push a little more and ask what feelings she had when making the pietà. She tells me, "I felt unsure. Well, erase that. If I'd been in my head, I would have wondered if people would think it's a fitting tribute. But I wasn't, it was just my gut. My gut doesn't give a shit what people think."

An every-Sunday-and-Wednesday Methodist until her thirties, she wandered through a variety of Protestant denominations until finding a comfortable space in the Episcopal Church. But she defines herself not as a Christian, no, but as "a musician in the Episcopal church." She is steeped in Christianity. And she rejects it. When her son died, she named her mourning piece after an iconic Christian subject yet used an object with roots in Minoan Greece to create her art. I am torn about putting this experience of Mom's into words. Words often trap her. She is most free when she is outside the world of words.

I pick morels and lovage for today's lunch. Seven honeycombed cones, carefully staked out along the spring trail over the last week, await my knife. The lovage doesn't need staking; its wild abundance overflows the edge of the stream.

Full basket in hand, I return toward the house, past the sawmill where Lou Jean, Dad, and a new family friend, Steven N., are preparing yew logs for a potential buyer who makes longbows. I hear the chainsaw and wonder whether it's Dad or Steven N. I leave the giant sequoias lining the driveway and see Dad watching. Tears well up in my eyes. Steven N. is healthier than Steve was in his last years. He can use a saw. Dad can have a break, even if he doesn't want one. He attends to Dad. "I've got a pencil in my pocket," he says and writes the diameter Dad has measured out. Dad lets him.

A few weeks ago for Dad's birthday we all gathered in Portland for a Bach's Mass in B Minor at Trinity Cathedral. Lou Jean was anxious and emailed:

Dear Family,

I'm wanting to check in with each of you about the upcoming birthday/concert on the 13th. If it's uncomfortable or awkward in any way for my friend to join us, please let me know. Admittedly, I'm feeling a little apprehensive. So, it's not a problem at all for him to not come. I just want us all to enjoy this family celebration.

Love, LJ

Dad replied: "Lou Jean, I am fine with Steven being in the group. You are family, your friends are family. Doug."

Steven N.'s presence makes me think of Steve at every turn. Emotions roil and exhaust. He wears logger jeans and suspenders. He was a timber faller for many years until the vibration of the saw numbed his nerves, leaving him with "white finger."

He and Lou Jean were classmates from kindergarten through twelfth grade in Toledo. They drank the same raw milk from Oberg's Dairy; fat cream globules made their child selves grimace. Lou Jean worked at Cherry Lane Egg Farm. So did Steven N.'s mother. Neither ate eggs for years after.

I come into the house and tell Mom how his presence makes me feel Steve in each rock and bird, and flower and tree. I tell her it feels bittersweet. She agrees and says, "I am so happy Lou Jean has found him. I know she's been lonely. She was probably lonely with Steve. Well, we're all lonely in our marriages. It's part of the human condition."

I boil potatoes for a soft bed to soak up the garlic butter mushroom goodness.

*2013. Eight miles from where we grew up, along roads lined with second-*growth forests, is my brother's house. Steve is still with us, and we are gathering the winter's firewood supply. Dad, my fourteen-year-old nephew, Griffin, and I pass through a metal gate. Welded on the gate's top bar is a Hopi man-in-the-maze spiral—Steve and Lou Jean's welcome to all wanderers in life. A gravel driveway winds along a walnut grove on the right and an acre of gangly teenage firs on the left. A barn houses Steve's cedar-strip rowboat *The Sweet Lou Jean*, chainsaws, axes, mauls, wedges, orange logging tape, the smell of oil, sawdust, and sunlight. Adjoining the barn, a woodshed opens to the air, a deep gray sky thick with stacked clouds. A gas-powered splitter stands silent but will soon necessitate earplugs and preclude any conversations beyond requests mouthed widely, accompanied by hand signals.

Steve greets us, a purple ring lining the bridge of his nose—left over from the morning's beet-powered, chemo-counteracting smoothie. He and Dad are identically dressed: hickory shirts, suspenders, Wild Ass Double Logger Pants, and American Forest Foundation hats. As we power up the splitter, gas exhaust flows around my face. Steve is on his knees wrestling the blocks into position under the blade. I pull the lever down just in time to split a piece off the new round of wood. It's a game to get the timing just right. Dad lines up the next blocks and gathers fallen-off bark to mulch young trees, and Griffin throws newly split firewood on top of the pile, his young limbs strong and easy.

Griffin and I trade places; I want to move my body, to feel my muscles and bones. I heft the still-wet, pitchy pieces into perfect empty places. I throw and throw, trading arms. I throw higher than Dad's prescribed measure of four feet. Steve wants the pile six feet high. He's made a marker on a peeled pole to show Griffin. Dad shakes his head in disapproval. Steve wants the woodshed full to the brim, full of fuel for him and Lou Jean to feed the furnace through the rainy winter days. I throw as high as I can, wanting wholeness for Steve.

There is violence in this throwing—the thrown wood vehemently burying father-son disappointments, unspoken disappointments laid bare by old age, by the grief of illness striking too young. Thrown wood takes the place of words needed for healing, for soothing. Words lie stacked in the pile like the bones of soldiers built into the Great Wall of China. With each throw I fling pain, hoping to alleviate Steve's hurting. I throw blackness—thick, blank, deep, unknowable blackness—throw it away in hope of light. I throw innocence, youth, willful ignorance of family imperfections. A large piece of fir rolls off the pile, crashes into my right shin, and starts a bruise I can use as a worry stone.

I roll a block with my foot; bones in my lower spine shift and pinch and make a lasting ache, a place to work through the devastation, the family earthquake that is my brother's cancer. Wood falls from my grasp, crushes my left toe, and cracks my spa pedicure. The crack, exposed when I'm back in my city sandals, will remind me to intone my brother's name. I am strong in the woodshed: I push back, and in the midst of anger and confusion, I create something.

1976. "Rise and shine, glad to be alive and in the Corps!" Dad's voice wakes his three teenage children at seven on a Saturday morning in

May. Gloves, hat, red rubber boots, rain gear. With a few groans we slide onto the truck tailgate, peanut butter toast in hand. Dad drives us a gravel half-mile into our woods to a maple tree he's felled, limbed, and bucked into rounds. Dad and Steve split the rounds with axes and wedges. Julie and I bend down, hoist two or three pieces, cradle them against our chests, and trudge through sword ferns and nettles to toss the wood in the back of the pickup, our multiple trips making a muddy path.

At noon, Mom brings a thermos of Campbell's tomato soup and tuna sandwiches on homemade bread. We sit on our rain gear under the heavy limbs of a fir, chatting about Steve's basketball game the night before. Back at the peeled-pole woodshed, Dad and Steve throw the wood from the pickup, and Julie and I carefully stack it—the corners with two pieces running north–south, then two pieces running east–west. Our gloved hands work, pulling and stacking up to just four feet, each two rows creating a prescribed cord: four feet tall, four feet wide, and eight feet long.

The woodshed is full, full of promised warmth and light. Each night in fall and winter, Dad requires a fire to be lit in the fireplace by the time he gets home. His tires crunch on the driveway. Steve wads the *Longview Daily News* and places it on last night's ashes. Julie runs for the cedar kindling and crisscrosses four pieces on top of the paper. I fill the wheelbarrow from the woodshed and bring in three pieces—a heavy one for the base and two smaller ones to create the optimal triangular, oxygen-giving shape. A single strike on the Diamond Match box lights the fire. The door opens, and Dad walks in. The flames welcome him.

2018. I walk the road from the house to the clear-cut where Dad and Lou Jean light slash piles. Reddish-brown sprays of cedar leaves splotch the gravel. Rumbles from the nearby gravel pit roll up and mix with the croaks of frogs. The underbrush wears its early fall costume; the un-picked fruit of the salmonberry hangs black and soggy, but the green leaves thrive. Thwarted honeysuckle vines swing, clipped from their roots. Sunlight filters through the clouds and sharpens the shadow of my pen as I write in my notebook.

Two sets of "piss cans" settled on stumps welcome me to the clear-cut. These are five-gallon metal backpacks with a hand pump to put out any spreading fire. Above the pluming smoke loom the snags we've left for birds of prey. Purple thistles, slimy and dark with rain, stand next to the water trailer. My back is out, a middle-aged problem I have no patience for, and I can't climb over the rough ground to light the piles.

Ashes flutter down as I watch Dad walk between piles, a drip torch balanced on his right thigh. Billows of white smoke. Red-orange flames. Yellow pitchy smells. Green blackberry leaves. Black-clawed branches crumble and fall. A stack implodes. Lou Jean lights a new pile, diesel and smoke perfuming her purple-and-plaid work shirt.

The smoke makes shifting shadows. A jet drones over. A flock of geese honk, their formation lengthening and then shortening as they fly northwest.

The 110-year-old trees no longer stand in our forest, but they are not altogether gone. The logs are traveling to mills near and far to be reshaped, repurposed, transformed. Two-by-fours, temple beams, decking, plywood—the carbon absorbed over the century remains in the wood wherever it is. The wood carries the sun and air from the ridge above the Cowlitz River and shares it with the carpenters and house dwellers for years to come.

Peter Sr. piled the leftovers of J-logs (logs exported to Japan), China logs, sawlogs, and pulp into high haystacks for quick burning. Branches, treetops smaller than five inches, and log ends lie scattered across the ground and will feed new trees, but we still have enough slash for a hundred piles across our twelve acres.

I don't like the burning. It feels like a waste of good mulch or habitat. I've learned that some forest owners have stopped burning— they pay the planting crews more to plant through a thicker layer of branches. I will talk Dad into doing that after our next harvest. But for now, I comfort myself: our trees have scrubbed the air of carbon dioxide and sequestered it for sixty years longer than the industrial average of forty years.

Memories from a past burn, the first I was in charge of, flutter in my mind. We'd just finished a harvest at Callison the summer Steve passed. Twenty-three acres of slash piles needed lighting. Dad had a hiking trip planned for Nepal in early October and wanted to get the piles burned before he left. But rain didn't come and Lewis County didn't lift the burn ban until three days before his departure.

The burn permit came via email, and I sat at the tree farm's kitchen table reading it over.

State of Washington Department of Natural Resources
Burning Permit
No ignition before 7 am or after 4 pm.

"No problem—good hours for me."

Maximum number of piles to be actively burning at once: 10.

"Too much wiggle room in this rule—Dad is going to try to burn more at once." My stomach started to hurt.

Fires must be completely extinguished within 7 days of ignition.

"What! Slash piles take that long to burn? Dad will be in Nepal. I can't do this!"

But Dad wanted to start. "Burning is all about timing. The rains will start soon. Soggy piles are a pain in the butt."

I argued, "But the permit says only ten piles a day. We won't even be halfway through by the time you leave."

He shook his head. "Those rules aren't really cast in concrete."

I had the habit of following rules, but not the habit of refusing Dad. My mind and body whirred as we prepared for the burn. We used the shop hose to fill the water tank; we put the four-wheeler on its trailer; we gathered shovels and hoedaggers. But as Dad drove back to the house, I turned to him. "I can't be in charge of this. I don't know enough about fire. We have to wait until you get back."

Dad stopped the truck, his eyes narrowed and his jaw tightened. "Well, you're in charge."

As he packed for his trip, he was silent and dark. I barely withstood his disapproval.

While Dad was in Nepal,

      eight

          inches

              of rain

                  soaked

                      our

                          slash piles.

But that was five years ago, and I've figured out how to better hold my ground with Dad. The piercing pain in my back has subsided, and today I am determined to light some fires. We wait for the fog to lift so the smoke will rise straight and not head to the neighbors. Dad ignites his torch with a match, then Lou Jean's. "Let there be light."

The diesel drip torches are heavy and Lou Jean brings one close so I can light a pile. I'm wearing the hickory shirt I wore during the Callison burn—at the right shoulder is an ember-scorched hole Mom mended. I find a perfect spot, a small cave protected from the rain. I lift the bottom of the torch, and diesel passes through a metal loop into a flame and sets fire to a dried cedar branch. The crackle and the pungent smell remind me of throwing dried Christmas tree branches into a fire. Dad says fire is a fascinating critter. I agree.

Two piles lit, I sit on a stump to spare my sore back. I've forgotten my notebook and I'm writing on the back of the Department of Natural Resources Burning Permit. A ladybug rests on a sword fern the summer heat has curled. Deer droppings feed nettles, and new leaves have sprouted at the bottom of wilted stalks. This forest is never truly dormant.

I think about the word "ash" and dislike its broad meaning; I'm using it now to describe the residue of burning wood, and I used the same short syllable to describe the residue of Steve's body that we spread over the land. I wonder whether Japanese has different words for these. I look up the word "ash" in the Japanese dictionary on my phone. And it does: *kotsubai* for bone ash, *kibai* for wood ash.

The water tank isn't working well, and Dad wants to empty it, unplug the filter, and fill it again. We hook the pump trailer to the pickup, drive to a pile, unroll the hose, and Dad sprays a couple of smoldering circles. I get tired of watching, so I take a shovel and pile dirt on the edge of a huge glowing root. Burning pitch creates an aromatic friend as I play around making the smoke rise in different directions depending on how I throw the dirt. Dad walks over and says, "We don't need to put dirt on it if we're going to water it." I nod. Dad drives up to the next pile and asks me to signal to him where to stop. I do, but he's not satisfied and says, "We'd be better off with the trailer here," indicating thirty feet farther up the road, "with the hose half-assed lined up." I nod, but annoyance rises with his need to fine-tune my every move.

He's spraying another pile, and I lean my shovel against the truck and scrounge some paper from the glove compartment and take Dad's reliable pocket pen. As I write, Dad adjusts something on the hose and gets a good spray of water. He sees a pile with blue smoke a little way from the road and says, "I'm going to check on one over here. You wait." I walk toward the pile to get away from the noise of the pump. He notices and shakes his head disapprovingly. Instead of nodding, I blow him kisses. The pump is making too much noise to say, "Dad, I have my own brain! Back off." He blows kisses back.

Dad invites Lou Jean and me up to a bench overlooking the clear-cut. A dead cedar tree, seven years old, stands brown and dry at the beginning of the trail to the bench. Dad points it out and says, "I've been coveting that baby. But it's up on a small hump and didn't keep enough moisture. There are two other cedar behind these pine. I wanted three." We sit on the bench and listen to the crackling. Dad points to the nearest pile, now just ashes. "It's the only white pile—it was all alder branches—I've never noticed that before. It was a little hard to touch off, but it has cooked down great."

As we drive back to the house, we pass the Snag Patch, home of the saw-whet owl. Dad says, "Steve and I really worked hard to match trees to the soil in that patch. If we cut cedar, we planted cedar. In root-rot pockets we planted pine and in the other places we planted fir. It's really looking good. Maybe we can have a raven deliver a message to Steve one of these days." Lou Jean smiles and says, "The other day I was walking in the forest behind my house and a raven circled around me three times. You never know."

Unlike during the year of the Callison burn, rain doesn't fall for days. The piles continue to smoke. I arrive from Portland and Mom and Dad are gone. I'm here alone. I take a walk past the spring, through the young pines and the mature firs and cedars. The late autumn sun warms my back. I sit on the bench and look over the clear-cut. I see flame. After ten days? I walk down the hill and into the cut. I pick up a branch, thinking maybe I can knock over a burning root. But as I get closer, I see that it is a stump on fire. Dad's told us that in the old days of broadcast burns, when whole units were set on fire, foresters had to watch out for burned-out stumps. Old-growth stumps five, six feet across burned into piles of cool ash, but coals still smoldered deep into their root systems, posing a danger for anyone walking about. I'm glad we have only one small stump burning. I need a shovel or a piss can to put it out, but I decide to let Dad deal with it.

I walk back to the house and work in the garden. I feel calm and useful as I put the vegetable beds to sleep—pull up spent cucumber, tomato, and pepper plants, turn the soil and cover it with oak leaves, its blanket for the winter. The red of the leaves and the gray of the cedar-sided beds play off each other.

Dad gets home and I tell him about the burning stump. He drives toward the clear-cut in the pickup. A minute later, he backs up out of the woods, grinning, and says, "Hop in, I've got something to show you." I get in, silent; I know he likes the suspense. He says it's white and gray, on the left side of the road at about eye level. I ask, "Is it an owl?" He nods and says, "I hope it's still there." It is.

I roll down my window, sit up on the door, and take photos. We creep closer and can see a large barred owl turn its head, its gray-and-white-patterned feathers checking like an Escher drawing. It turns back to look at us, blinking as if to say, "Time for you to go now." But we want to see it fly. Finally, it swoops down, glides down the road about ten feet, and wings back up into the woods. "That was cool!" Dad is so happy. I am so happy. It is a great moment to share. I walk back to the house to finish putting leaves on the vegetable beds and Dad goes to take care of the burning stump.

Later, Dad and I walk around checking piles. A red-tailed hawk sits on the tallest snag. Dad smiles. "He's saying, 'Thanks for leaving this nice perch for me.'" The hawk flies off and the late afternoon sun catches its tail feathers. Dad points. "That color always makes me think of your mother's hair when she was young—it was a beautiful auburn color."

We look at one last pile, a circle of foot-deep ashes with a four-foot-high charred root on the edge. Dad says, "The rains will tamp these ashes down. Those roots will last a century at least, providing good shade for a new tree."

Now just a few wisps of smoke float above the slash piles and we can start tending another stand of trees. We work above the clear-cut where we can keep an eye out. Ponderosa pine, fir, and cedar struggle to thrive in this rocky patch of soil. Dad and I bring discarded paper and cardboard from the oil shed to mulch the eight-year-old trees.

The Douglas fir are sick. The seedlings thrived their first six years; two-foot leaders reached to the sky every spring, fluffy green-tipped branches stretched out to their neighbors. But this year, this spring after four long summers of heat and rainless skies, the stressed Douglas fir are losing their needles, and small cankers dot their limbs. Three years ago, I noticed a few red branches and asked a forest pathologist from the Department of Natural Resources to come. She placed cut samples into Ziploc bags and sent them to a lab at Oregon State University. The lab emailed me:

The insect pupa from the Douglas-fir twig was a weevil pupa. Because this pupa was fairly well-developed, it showed a lot of characteristics of an adult weevil. Notable characteristics were the large eyes that nearly meet on the front of the face and that were situated high on the head, dorsal to the base of the rostrum. It also had relatively short antennae attached near the base of the rostrum. These characteristics place this weevil in the subfamily Conoderinae. Although it is not possible to key weevil pupae to species, the fact that it came from a Douglas-fir twig, and that it is in the subfamily Conoderinae, almost certainly make this the pupa of a Douglas-fir twig weevil, *Cylindrocopturus furnissi*.

Management Suggestions: Keep trees healthy. Remove and burn infested twigs. Avoid planting on drought-prone or wet sites.

The management suggestions feel like advice for a suburban backyard, not a working forest. Some of our infested twigs are twenty feet high. We have Douglas fir planted in each of our three hundred acres. How do we keep trees healthy when there is no rain? When the weather patterns do not support the tree species that have grown on this ridge for a few thousand years? It's no longer a few trees. Almost every Douglas fir tree we've planted in the last twenty years has at least one red tip.

On one tree after another, limbs turn red, needles fall. Naked black twigs remain, stark against the deep green of new grass and salmonberry leaves.

Dad and I leave the fir to their fate and place paper and cardboard around the pine and cedar. It feels heartless. My usual joy and satisfaction of working in the woods has given way to existential angst.

Dad's shoulders slump. Our booted feet move heavily over the uneven ground. Our forest has always been a hopeful place, burgeoning and bursting. Dad says the weevil threat reminds him how farmers must have felt during the Dust Bowl. "Their mouths were too dry to lick a stamp to write and ask for help."

I come back to the house and email the forest pathologist, hoping she's learned something new. This answer is more disconcerting than the first:

There is the tip weevil, the Douglas-fir pole beetle and two fungal canker diseases causing top and/or tip dieback on Douglas Fir trees. The two bugs and two diseases are normally insignificant, but with the recent hot dry summers, and even winters that are warmer and drier, these things are becoming more common and even causing mortality in certain cases. Because these are normally inconsequential, we have no management recommendations for them. We're likely to see more of this as long as drought conditions persist.

No management recommendations?

A little wingless bug can devastate our trees, and we and the trees are without defense. We can't set traps baited with rock salt and apples. We can't come stalking with a shotgun in the middle of the night. We can't protect the fir with Vexar tubes. We can't sever the weevil with our clippers.

Humans live in nature; we must protect it so it will sustain us. But so far, the sum of our actions—too much taking without replacing, polluting air and water with the spoils of industry—has resulted in a turbulent, extreme nature.

It is early September 2020. This year abundant rains soak the ground and the fir are flourishing again. But the rains have not stopped wildfires. The air at the tree farm is heavy with the smoke of burning forests, grasslands, and towns, all raging miles away. Acrid and thick, it feels apocalyptic. Mom wears a face mask to weed her heather garden; in just an hour her mask has turned dingy gray. Multiple fires, both lightning caused and human caused, have exploded with the arrival of extreme winds. Before the season is over, more than 1,713,000 acres and hundreds of homes in Washington and Oregon will burn.

Fire has always been part of the forest ecosystem. Tree ring records, early land surveys showing burned areas, and legends point to large fires over the centuries. The setting of this Pacific Northwest folktale about the Douglas fir cone, often but never definitively attributed to Native American sources and here retold here by Kevin Ebi, is a fast-moving fire, much like the ones of 2020.

It was the middle of the night, but the forest was brightly lit as if it were day. A lightning strike had triggered a massive wildfire.

There had been other lightning strikes and other fires, but everyone from the smallest creatures to the biggest trees knew this one was different. Nothing was slowing it down. Not only did it continue to grow, but it grew at a faster and faster rate.

All the animals began to flee. Birds flew away. Deer and many other animals ran, barely able to beat the heat. But the mice, with their tiny legs, fell farther and farther behind.

The mice ran as fast as they could but the fire was nipping at their tails. The heat grew unbearable. They frantically looked for shelter. They came to a tall big-leaf maple tree—one of the tallest in the forest.

"Can you help us?" they asked the maple tree.

"I'm afraid not," it replied. "I'm worried for my own safety. I can't stop rustling my giant leaves with this wind. My leaves are only fanning the flames more."

Next, they came to a western red cedar.

"Can you provide us shelter?"

"I'm afraid not," the cedar responded. "My bark offers no protection from these flames. It's already starting to singe."

Finally, they reached a Douglas fir, who urged them to climb it to safety.

"Hurry," the Douglas fir said. "My bark resists fire. Climb me to get above the flames."

But it had been a very dry year, which provided the fire extra fuel. Many of its flames were as tall as the trees.

"Fear not," the Douglas fir told the mice. "Climb inside my cones. They will protect you."

The flames were only a fraction of a second behind, however. The mice dove headfirst into the nearest cone, making it only about halfway in before the fir tree had to snap the cones shut to provide them safety.

To this day, when you look at a Douglas fir cone, you can still see the hind legs and tails of mice sticking out, still seeking shelter from fire.

But forest fires have increased in severity and frequency. Growing numbers of humans live on land that was once forest. Unlike the mice, we cannot hide in the Douglas fir cones. And unlike the mice, we are not blameless. The side effects of our comfortable, industrialized lives destabilize the climate. Prolonged, increasingly arid Pacific Northwest summers make forests more susceptible to pathogens and pests. While the mighty Douglas fir has long dominated our local forests, root rot, fir twig weevils, and *Phomopsis* threaten it. The trees these kill are perfect fuel for fire. The lack of rain stresses the animals, too. Our forest deer drink from our streams and wetlands, but in the summer, they have to trek over the ridge to the Cowlitz River. And as the summers stretch on, they must do this more and more.

So far, fire has not threatened our forest. Our tree farm's topography protects us from the most extreme wind. Still, each summer we are on high alert. At a whiff of smoke we'll jump in the pickup to patrol the gravel roads for errant embers from a neighbor's ill-advised brush pile. Or we'll just cruise the country road that divides our farm, on the lookout for tossed cigarette butts. A smoke jumper in his early twenties, Dad has a keen sense for shifts in humidity and wind that can redirect fire in an instant. Each August he and I fill our piss cans and set two in the back of the pickup and two on wood blocks along the driveway. When full, the metal packs weigh fifty pounds and I can't put one on from ground level. We fill our three-hundred-gallon pump trailer and position it to use in an instant. And we pray for rain and soft winds.

The tree farm is

70 miles north of Portland, 100 miles south of Seattle
  Exit 57—follow Jackson Hwy north, past the barn bearing
  the old advertisement: "Dr Pierce's Favorite Prescription For
  Weak Women Makes Red Blood." Right on Hwy 505 toward
  Mount Saint Helens. Left on Eadon Road. Left on Collins
  Road. 519 Collins Road has been my parents' address for 47
  years
20 miles south of Chehalis, scene of my first date to see
  Saturday Night Fever
3 miles east of Toledo, WA, population 600
15 miles from the nearest Starbucks
5 miles into southern Lewis County, the only Washington

county west of the Cascades that didn't vote for Obama in 2008

40 miles west of Mount Saint Helens. My brother's high school graduation was postponed because of the eruption

2,987 miles from my NYC apartment home of ten years

1 mile as the crow flies from Saint Francis Xavier Mission, established in 1838. The first Catholic mission in Oregon Territory

30 miles from the Port of Longview, home of Weyerhaeuser and Port Blakely log yards

4,885 miles from the west coast of Japan and 5,269 miles from the west coast of South Korea, both destinations for our logs

The logging site is

1,000 feet from the house where I grew up

1,000 feet from the dormant vegetable garden, the empty doghouse, Mom's ceramics studio, Dad's shop, my new fruit orchard, the woodshed

1,400 feet from the log cabin, built by Steve in the 1980s with trees from our forest

1,500 feet from The Swing hanging from a massive big-leaf maple—the tree whose branches graced my sister Julie's wedding. That tree is where I attempted to smoke my first cigarette. And even today it's a gathering spot for all young visitors wanting to be out of their parent's eyeshot

2,000 feet from the one log sawmill that has hosted numerous parties, my wedding to Tom, and Steve's Celebration of Life

PART TWO

Phloem

The inner bark, or "phloem," is the pipeline through which food is passed to the rest of the tree. It lives for only a short time, then dies and turns to cork to become part of the protective outer bark.

The burning is over, and I decide to go back to the Snag Patch to find more honeysuckle. I find Dad in the shop and ask him to sharpen my clippers. As he works, I look at a sign that's hung on the wall for forty years: "Do Somethin' Either Lead Follow or Get the Hell Out of the Way." Across the concrete-floored room stands a log where Steve's timber-falling saws hang—twenty-pound power heads with thirty-six-inch-long blades. The saws remind me of the tension between Dad and Steve. I think about Steve's desire to lead and Dad's inability to follow. I am angry at both of them—why couldn't Dad let go? Why didn't Steve grab control? Why the awful battle of wills?

Dad's weaknesses are part of my family's narrative, especially Mom's version. But Mom does not see, willfully does not see weaknesses on the part of their three children. Dad says that Steve is her "golden boy." He says often that if one of us ran onto the freeway, he'd look both ways and run out to save us, but that Mom would run out without looking. He says this with rueful pride. I think he wonders whether she would run out to save *him* without looking. Trees that grow too close together compete for the same nutrients from the soil and sun. Mom was a main source of nutrients for Dad and Steve, and their competition was fierce.

In Japan there is a saying: "Before you get married, look with both eyes. After you get married, close one." Mom has looked at the three of us with one eye closed, but always at Dad with both eyes open. As I write and try to look with both eyes at Steve, or Julie, or myself, I feel as if I am betraying Mom.

Ditch Flowers

The impact dislocates his hip.
Steve's head cracks the windshield.
A concrete culvert.
A sharp turn on a country road.
2002. A dark night. Too many beers.

Foxglove: pink, purple bell-shaped flowers on tall stalks. Dad smiles when he tells a story of seeing an elk plunge through a new tree plantation that was sharing space with hundreds of foxglove. Myriad flowers swirled up behind the running animal.

A DUI is not an option.
A phone call from Steve
brings Lou Jean

before the police arrive.
She and a friend
transfer Steve to her pickup
He lies on the truck seat
for hours
before going to the hospital.

Thimbleberry: my favorite berry for eating right in the woods. Velvety thin red flesh hugs a rounded core. A perfect fruit for picking on a walk with a loved one.

Mom and I bring Steve
some of Dad's clothes in the hospital.
His have been cut off him,
Dad's shorts don't fit.
Steve snarls, "God damn skinny son of a bitch"
and throws them on the floor.
I'm taken aback by his anger.

Scotch broom: a noxious weed, an invasive species. Will rob new trees of all soil and sun if not controlled. Brilliant yellow flowers hide its scourge. Its seeds rest ready in the soil for over fifty years.

Dad is proud of his physique.
Yoga every day for years.
Working in the woods.
Denying his sweet tooth.
Dad calls Steve's type of belly
a Heidelberg tumor.
Never to him or about him
just derisively
in his presence.

Red huckleberry: Mom and Dad paid us $0.25 a quart for these quarter-inch berries. A lower wage than the $1.50/hour for mulching trees.

Steve leaves the hospital,
Lou Jean sets him up
in Mom and Dad's living room.
He lies in his hospital bed
next to the rug
where Dad does his daily yoga.

Dad and I hike around Coldwater Lake by Mount Saint Helens. His furious strides fling up dust in front of me. His words fly back, "If booze is such an issue for him . . . he might not be the right guy to run the tree farm. "

"How can you think that? This will wake him up."

Dad's shoulders tense and he shakes his head. We are silent as we continue along the clear blue water and look up into the mountain's caldera.

Each Wednesday evening during middle school, Mom dropped Steve, Julie, and me off at the Longview Public Library while she led choir practice at St. Stephen's Episcopal Church. For an hour and a half, we had free range of the stacks, magazine racks, and card catalogs. The library limited each child to ten books; we always used our limit. I have one memory of driving home in the dark and asking Mom to stay right in front of a car so we could use its headlights to read.

Bags and boxes of Steve's books sit in our garage. In the weeks after his death, Lou Jean needed them away but not gone. Four years later, they are still here, unsorted, the bags and boxes sagging. I go downstairs and look, wanting to add some layers to Steve's life in these pages.

I expect to find books by Wendell Berry, but I don't. I call Lou Jean; she has kept them and brings them the next time we meet. A small stack sits on my table: *The Unsettling of America / Life Is a Miracle / It All Turns on Affection / Sex, Economy, Freedom and Community.* Between the pages of the last lies a handmade bookmark with a quote in Steve's handwriting: "If we do not live where we work, when we work we are wasting our lives and our work, too."

Steve's Books	Ideas in Action
Maintaining Biodiversity in Forest Ecosystems / *The New Economy of Nature* / *Wildlife Habitat Relationships in Forest Ecosystems* / *Managing a Non-profit in the 21st Century* / *Planning Family Forests* / *Old Growth in A New World*	In 2002, Steve started the Family Forest Foundation "to promote the conservation and sustainable management of family forests." One of his first projects was a collaboration with US Fish and Wildlife Service biologists to create a Habitat Conservation Plan for Lewis County. After ten years, multiple visits to DC, and hundreds of thousands of dollars of federal grant money, the plan was rejected by the USFWS. It was a bitter blow. I am the current president of the Family Forest Foundation; we are working to continue Steve's vision.
Dog Owner's Home Veterinary Handbook / *How to Raise a Bouvier*	Big, black, and furry: Bodean, Zack, Zeppelin, and Izzy lived with Steve and Lou Jean in their homes, ate special food, and went to the veterinarian. Mom and Dad allow Izzy into our house now.
Drift Boats: A Complete Guide / *Drift Boats and River Dories: Their History, Design, Construction, and Use*	Steve named his boat *The Sweet Lou Jean*. He rigged it up so they could catch crab. In Steve's will, the discussion of who should inherit *The Sweet Lou Jean* takes more paragraphs than any other asset.
Cancer: Step Outside the Box / *Integrative Oncology* / *The End of Pain* / *Outsmart Your Cancer* / *MycoMedicinals: An Informational Treatise on Mushrooms*	Steve tried everything.
Flour, Water, Salt, Yeast: The Fundamentals of Artisan Breads and Pizzas / *Forks over Knives* / *Food-Healing: Cooking with Qi*	When Steve knew a meeting might be contentious, he made muffins. He brought tree-planting crews his chocolate chip cookies. In 2012 he built a pizza oven at Mom and Dad's and delved deep into the chemistry of pizza and bread dough.
100 Hikes in the Pacific Northwest / *Paddling Pacific Northwest Whitewater* / *How to Rent a Fire Lookout in the Pacific NW*	Steve, Lou Jean, Dad, and Julie paddled the Cowlitz, the Nahanni, the Deschutes, the John Day, the Owyhee, packing the canoes for week-long explorations.

When we were young, in the pre-earbud 1970s, Steve's music choices ruled our living room soundwaves. I had just begun exploring Baptist Christianity, and my first purchases included albums by Second Chapter of Acts and Keith Green.

Steve's CDs	Titles/Lyrics	Lou Jean's Words
Rockin' Razorbacks	More Love, Less Attitude	We saw them many times at the White Eagle. It was my introduction to Steve's music world.
Stevie Ray Vaughn	You're My Pride and Joy	Joe Blatt played that for us at Steve's fiftieth birthday party.
Too Slim and the Tailgaters	Tales of Sin and Redemption	They were a local Portland group that caught Steve's attention. We saw them at the Blues Fest and a couple of small venues.
Jason and the Scorchers	Clear, Impetuous Morning	Talk about headbangers! They were fun. Of course, you know, we were much younger then.
Sleater-Kinney	The Woods	I didn't like them much. You went with Steve to hear them once.
Lynyrd Skynyrd	Call Me the Breeze	That's what he was playing when you guys were in high school.
Beth Hart	Leave the Light on	That girl can sing!
The Paladins	Lover's Rock	Steve proposed to me on a rock on Steamboat Island.
Sue Foley	The Forest, Queen Bee	We saw her with you and Tom at the 2013 Winthrop Blues Festival.
Johny Winter	Life Is Hard/ Nothin' but the Blues	He was so raw. I could see why Steve liked him.

Songs such as "To Obey is Better Than Sacrifice," "Create in Me a Clean Heart," or "Humble Yourself" had no chance against AC/DC's "Dirty Deeds Done Dirt Cheap," "Love at First Feel," or "Highway to Hell."

I ask Mom whether she remembers Steve's first concert. She replies, "No, I probably didn't know about it."

Steve's CD collection is in Lou Jean's light-filled living room. I sit on the couch and ask Lou Jean to talk to me about his music.

2012. After Steve's chemotherapy and surgery, he drastically changes his diet: gluten-free, dairy-free, low carb, lots of fresh vegetables and greens, exercise. During that time, Dad makes a written pact with Steve not to drink for a year. He and Steve sign it, and Lou Jean, Tom, Mom, and I add our signatures as witnesses.

> "Upon assessing our current physical conditions and common future goals, Steve and Doug Stinson are committing to abstaining from the intake of alcoholic beverages for one year beginning Sunday, September 16, 2012, and ending on September 16, 2013."

Neither drinks in each other's presence for the year.

Months later, Lou Jean and I drink gin and tonics on the deck. Shaking her head, she says, "Steve didn't keep to that contract. I think he resented your dad for it."

Dad finds this agreement as he is cleaning his desk out along with his mother's journals, a letter from me in college, photos of him as a chubby, black-curled toddler and as a stern, buzz-cut marine. He sighs and puts it in the recycling bag. I retrieve it and bring it home for my writing project.

I show the contract to Tom and ask what he thinks about it and Steve's resentment. It reminds Tom of a contract he wrote with his siblings to clean the house in the wake of his father's anger. But Tom was that contract's author, and he was eight years old, not fifty-two and dying. I understand Steve's indignation. But at the bottom of the contract is my signature. I was following.

Fragmented Memories

Nineteen-year-old Steve is prone next to the dining room table. He's been smoking pot with something crystal-like on the joint. It's made him sick with dry heaves. I feel alone and helpless.

A few months before Steve passes, he, Dad, and I go to Callison to meet a log buyer. Though Steve has been co-manager for three years, he doesn't have his own business card. I write his name and phone number on the back of Dad's card and give it to the log buyer.

Dad tells me that his boss at RHD Plywood wanted to hire Steve, but Dad told him, "Steve looks like me, but he is his mother."

After a day in the woods, Steve asks Dad to light up the sauna. Afterward they stand next to the cedar fence, reddened bodies steaming as a light rain falls. Dad says the dark heat created an atmosphere for great discussion between him and Steve. The sauna sits cold and quiet these days.

At Dad's seventy-fifth birthday party Steve tells a story that's long been part of our family lore. "On my sixth birthday, Dad took me out into the woods to build a fire with just two matches and no paper. It was wet, but I found some dry needles and twigs. My first match fizzled. My hands shook." Steve's voice breaks. "The second match lit the needles and there was flame. My reward was my first ax."

I've always hated that story. The power balance is off; I imagine a small boy anxiously lighting a match and a large man powerfully watching. I wonder whether Dad hugged him when he was successful. I wonder whether Steve felt pride or relief or both. I wonder how much of my own desire to please Dad is the lens through which I see this story. I wish I could ask Steve what he feels about it.

Dad did not put his daughters to the same test. I did get my own ax at age ten, but for my birthday. I was not made to earn it.

This week, after I write the story of the first ax, Mom, Dad, Lou Jean, and I are sitting on the couch next to the wood stove. I ask Dad whether he hugged Steve after he successfully lit the fire. He shrugs ruefully and says, "Probably not." Mom grimaces and says, "It was just one of many tests Doug put Steve through." Lou Jean says Steve always told the story with pride.

At Steve's Celebration of Life, Dad said these words:

Born in Ketchikan, Alaska, December 10, 1961.

At ten days old, we flew to Edna Bay, ninety air miles northwest of Ketchikan, Steve's home for the next two years. A major snowstorm had hit southeastern Alaska the day before and now it was crystal-clear blue skies. We were on the first morning flight out of Ketchikan. A beautiful flight out—the mountains blanketed in fresh snow. When the Grumman Goose taxied up to the floating dock, the pontoon was resting in the snow. I carried Steve up to our house in three feet of snow. A good start for the young forester.

Steve's next stop was Roseburg, Oregon.

Age six: he earned his first cruising ax. He started his fire in the woods with only two matches. No paper.

Next stop, Washington State.

Age ten: we acquired the Toledo tract in 1971. Steve was soon planting trees, mulching, and burning slash. All the basics.

Age fifteen: Steve started cutting and selling firewood. His junior and senior summers, he worked at Elma Plywood pulling on the green chain, working with the millwright, whatever was asked of him.

Age eighteen: WSU at Pullman, majoring in forestry. He had a rocky start. At the beginning of his sophomore year, he called me and said, "Dad, I'm wasting my time and your money." Today, I think what Steve was saying was "I have done it your way for twenty years, now I am doing it my way."

That spring, he planted trees for Tom Fox, and that was the beginning of an everlasting friendship. From a fellow tree planter around this time, Steve acquired his famous apple pie recipe. He then became a timber cutter, a life he followed for ten years and truly loved.

1988: Steve and Lou Jean Clark met. You know the old adage "Behind every strong man is a stronger woman." Lou Jean makes me remember that dynamite comes in small packages.

1990: Lou Jean went to Evergreen and got a degree in environmental science. Steve decided to go to Evergreen as well and then on to the University of Washington for a master's degree in forestry. The fire was lit. From that moment on, Steve was passionate about good forest management.

You all know the rest of the story. Some highlights from Steve's achievements for the Cowlitz Ridge Tree Farm:

1. Snag retention: you saw it along Collins Road.
2. Conversion of a thirty-three-year-old Douglas fir stand, severely infected with root rot, into a ponderosa pine / western red cedar plantation.
3. Steve procured three alternate plans for our tree farm.
4. Steve finalized our fifteen-year plan with the DNR.

Steve's life was short, full, visionary, and generous.

Thank you, Steve.

PART THREE

Cambium

The cambium cell layer is the growing part of the trunk. It annually produces new bark and new wood in response to hormones that pass down through the phloem with food from the leaves. These hormones, called auxins, stimulate growth in cells. Auxins are produced by leaf buds at the ends of branches as soon as they start growing in the spring.

Mom and Dad purchased our land from Elmer and Dorothy Boone, a longtime Toledo family. They stayed in their home on the tree farm for the first several years after we built our house and were our surrogate grandparents, hosting us for Thanksgivings and sometimes babysitting. Elmer helped tape and sand the Sheetrock as our house took shape. He taught us about tansy, a noxious weed, and showed Mom how to roll under a barbed-wire fence. On our birthdays we received cards with slots for quarters and Dorothy made us birthday cakes using Duncan Hines chocolate cake mix and store-bought frosting. Our granola- and yogurt-leaning home never saw this sort of treat, and the sugary mix thrilled us. After a few years, Steve requested an apple pie rather than a cake. And so every December 10 an apple pie appeared. Once he ate the whole pie himself. After that, she made two; one for the family and one for Steve. Julie and I stayed with the cakes.

As children we told people we lived on "the old Boone place" and they knew exactly where it was. Elmer had purchased the acreage in 1936 from his mother, Lenora Boone; their name was long attached to the land along Collins Road. I wish Elmer and Dorothy were still alive so I could ask about their early days here and listen to stories about previous owners and uses of the land. Instead I must start my search with silent dusty paper.

I begin at the Washington State Archives in Olympia. I've asked Dad to accompany me; his fifty-year knowledge of our property will help me notice clues in the deeds we've come to examine. We park near the capitol building. Rain hits the streets and has to run for a few yards before it finds soil to sink into. Searching for the building, we hurry, trying to avoid a soaking. Rainy days are no problem on the tree farm. Rain gear from Service Saw and red rubber boots, sometimes called Ketchikan sneakers, allow entrance into the weather. But here in the city, we walk rapidly with our bodies hunched against the elements.

I've made an appointment with Jewell, an archivist, and she meets us at the door. She's gathered a cart full of deed records from Lewis County. Thick twelve-by-twenty-four-inch ledgers hold lists of names, dates, legal descriptions of lands, and many signatures. Hiding behind these lists are stories: dreams of a better future, a falling-out with a sibling, a father's death, a new marriage. I hope these stories make themselves known as our fingers run over the pages.

We start with grantee ledger B in search of Lenora Boone. The records show that Lenora bought the land (Southwest Quarter of the Southwest Quarter of Section 10, Township 11, North of Range 1 West of Willamette Meridian) from Augustus and Catherine Bonnin in 1904. The records become murkier after that.

We find a 1907 hand-drawn map and timber cruise. Various sections of the land are labeled in pencil: brush, field, partly logged, timber, pasture, barn, old channel of the Cowlitz River. The cruise determines there are 1,217 MBF (thousand board feet) of fir and 400 MBF of cedar. The section titled "Character of Land" describes the soil as "bench land: rocky." This information agrees with the oral history we have of our land: the old growth was cut between 1900 and 1910.

Four hours into our search, Dad is restless and Jewell is looking a little bedraggled. She tells us that most people quit their search after one cart. We've been through three. We leave with more names to look up, but no stories yet.

Back at home, my dining room table is scattered with deed records, the most recent ones typed, the older ones written in beautiful and difficult cursive. I find who sold the land to the Bonnins: a bachelor named Lewis Frey, who bought it at auction from Lewis County in 1902 and sold it to the Bonnins the same year. I cannot find how the county came into possession of the land.

But I do discover the name of the first person to receive an official US government deed to the land. In 1884, the Bureau of Land Management gave a patent to Edgar R. Willoughby under the 1862 Homestead Act. I dig deeper and find that Edgar was a Civil War veteran. The Washington Civil War Association has gathered many documents about his life, and they lead me deep into the lives of the first nonnative people to own our land. I have found a story full of turmoil, and the backdrop is our bench above the Cowlitz River.

Born in New York State, Edgar joined the Tenth New York Cavalry K in 1862 and remained a soldier until the end of the war. In 1863 he was treated for a head injury received in the Battle of Gettysburg. In 1875, at twenty-eight, he married fifteen-year-old Esther Partsh in Nebraska. Census records show five children, four of whom survived: Ella, Emma, Ethel, and Edgar Jr., the last three born in Washington. But the old head injury caused trouble and in 1885, one year after receiving the deed to the land, he filed an affidavit for an invalid soldier's pension. It describes what happened immediately following his injury in battle. He "was brought to [Mrs. Catherine Wentz's] father's house in a dying condition" and remained there "for a term of about three months at which time said Edgar R Willoughby was taken to the Hospital at Chambersburg, Pa." The affidavit was successful, and the resulting "Declaration for Original Invalid Pension" describes his job as a painter but notes that he is unable to work.

By 1900 Edgar and Esther were living separately about 120 miles apart. The 1900 Census lists Esther as head of the household and living on Eden Prairie near Toledo, Washington, with her three younger children, and indicates that Edgar lived in Everett, Washington, as a "live-in hand." And then a tragedy. A 1903 headline from the Everett *Daily Herald* reads: "Old Soldier Dies in the City Lockup: Was Found Intoxicated, and Suffered Collapse as the Result of Long Dissipation." The article continues:

> Edgar Willoughby, a veteran of the Civil War, died early this morning in the city lock up. Willoughby was picked up Saturday by the police and lodged in jail. When arrested he was in the rear of the Bank saloon. He went along with the officers with little difficulty and seemed in a very jovial mood. Yesterday his intoxication had worn off and he was rather despondent and complained of not feeling well.

The death certificate lists the cause of death as "heart failure, alcoholism, and asthma."

Esther applied to receive his soldier's pension. She certified "that during the last several years of soldier's life he was almost constantly away from home at work. That he returned at long intervals and sometimes sent me money. . . . Soldier was at Everett, Washington when he died and I was at Toledo, Wash with the family." Esther received the pension but died just seven years later at the age of fifty. Her death certificate lists the cause as "Myocarditis" and her place of residence as Toledo and indicates that she is buried in the town cemetery.

Why didn't the land sustain this family better? In 1889 Esther's name appears on a deed selling the land to Oliver and Mary Shead. Why did she sell the land after only five years? How long had Edgar been away at this point? Was she ever able to enjoy it? Derive comfort from the big trees? I wonder whether Esther was excited to homestead on this land above the Cowlitz River or whether the forest was too bewildering. Was Edgar's trial with alcohol already a problem on the trail from Nebraska?

I have a hard time finding Esther's headstone; I've never been to the Toledo Cemetery before. My grandparents are buried in Oklahoma and Missouri, and Steve's Tree holds his ashes close. I walk each row of the old sections, looking for the name Willoughby. I can't find it—the graves are not in straight lines and it is hard to keep track of where I've been. I see the headstones for the Boone family: Lenora and W. T., Elmer and Dorothy.

Finally, on my third visit, the groundskeeper stops his mower and asks who I'm looking for. "Just call this number," he says, and I type as he recites. A helpful woman answers and says Esther is buried in plot 11-4 in the old Kellogg section. The groundskeeper and I walk to that section. I tell him my name, and he says, "You had a brother that passed recently." I nod, surprised, but comforted to be known by a stranger. We find the grave: a square upright stone engraved with the words "Mother / Esther P Willoughby / August 27 1859 / Jan 28 1910." To the left of her grave, in the same plot, stands a headstone that reads "Martha Morris / Dec 19 1849 / May 12 1919." I wonder who she is. The groundskeeper looks to the sky and says, "We're going to get wet." I take a quick photo and leave just before a storm cloud bursts open.

Back home, I look up Martha Morris. It is Esther's sister! They are the daughters of John Partsh from Nebraska, whose headstone I saw near the sisters' plot. I am so glad that Esther had family around after Edgar went to Everett to work.

I ask Lou Jean to come back to the cemetery with me. Her parents are buried there. We drive through town, past the gas station and its sign advertising "Live Sand Shrimp and Microbrews," past the Assembly of God church and its sign announcing "Free Trip to Heaven, Details Inside," and left onto St Helens Street and into the Fir Lawn Funeral Home parking lot. Lou Jean shows me her parents' headstones and says, "They have a good spot here in the shade of the cedars." We visit Esther and Martha's plot. Martha's name is followed by the name of her husband, E. B. Morris, but only his birthdate, 1843, is engraved, not his death date. We are curious whether he is buried here, or whether both Martha and Esther are here without their husbands.

We sit on the concrete border, our shoes resting on the close-cropped clover and dandelions, and wonder aloud about the sisters' daily lives. I say that both are listed as "housekeeper" in census records and their men as "farmers," "shake weavers," and "laborers at clay factory."

I want to time travel and sit on a fallen tree together with Esther, feel the breeze, and talk. Would we understand each other's words, each other's souls? Can a connection to the land overcome the distance of time?

I choose a place in the woods where I think Esther may have come for water. Just one hundred yards from Collins Road, which shows up on the 1907 timber cruise map, is a spring. It gushes from the hillside and over the bluff. The spring provided Elmer and Dorothy Boone with clean, cold water for their family, and it supplies us now.

I sit next to a sixty-year-old cedar. Bird droppings stain its feathery bark white. A spider has made a web in the fluted hollows just above the roots. Unfurling sword fern and Oregon grape surround me, together with prehistoric-looking horsetails, still-tender nettles, and wild bleeding hearts (I like that lay botanists confirm human emotion with the names they bestow). Dead maple leaves crunch as I move my feet to get more comfortable. I pick a horsetail and put it in my pocket for Mom to confirm its identity.

A soft green wind floats up the bluff over the stream to my face. From my perch, I can hear but not see the water flowing over rocks, creating a more consistent sound than ocean waves. More consistent than the breeze. Steady. Clouds uncover the sun and rays filter through the cedar boughs and just-leafing maple. My hand is sticky from the horsetail. I ask Esther questions.

Did you ever drink from this spring?

Did you gather morels and lovage?

Did you ever see a saw-whet owl?

How did you support yourself when Edgar "was almost constantly away"?

I imagine her tired and worn down, hair in a bun, buttons running up the bosom of a cotton dress—like the stern photos of my great-grandmothers.

Did you enjoy being a mother? I do not have children. Is that unfathomable to you? Did you choose when your children were born? Choose how many? I don't even know whether my body would have borne children. I didn't give it a chance. My choices are a symptom of modernity; a separation from the fertile land. The conversation stays in my mind. I get up and walk back to the house.

On another day I invite Esther to join me at the spring. It's a hot July day and I change out of my shorts and tank top into thick jeans and hickory shirt so the underbrush doesn't scratch my skin. Esther arrives in her long skirts—I think she has probably never left her home with her arms or legs bared. She's brought her two youngest children, ten-year-old Emma and five-year-old Ethel. The girls help bridge the divide of time. My questions stay unasked; we play instead of talk. We decide to build a town under the cedar tree. Esther gathers rocks to make a road—her work-worn hands place them side by side. My softer hands snap the stems of ferns to plant along her road. I am happy to have company under this tree. As a child, about Emma's age, I spent hours building towns under the big maple near our house. My friends wanted to go exploring on their bikes, but I wanted to stay put. Emma finds fir cones to make a fence and Ethel pours springwater into a hole lined with salmonberry leaves. We find small cedar cones to float on the pond. I open my picnic basket and bring out sliced cucumbers picked this morning, last summer's pickled onions, boiled eggs, and five of Mom's ceramic cups. Esther's basket has homemade bread and butter and a berry pie. We sit next to our newly created town and listen to the perpetual spring spilling over the bluff. I pass Esther a cup full of cold water. I lift mine to hers and look her full in the face. I ask my eyes to let her know I will hold some of her pain for her. And I thank her for sharing her children with me.

In the early 1900s, loggers used specially shaped axes to cut notches, one on each side of a tree, about five feet above the ground. They then placed into the notches a two-foot-wide, six-foot-long "springboard" with a metal "shoe" attached, and standing on the board, they cut the tree down using a crosscut saw, also called a "misery whip."

Along the wetland trail in Gemini Grove, old-growth stumps still bear the notches of springboards, and once Dad found a springboard shoe and an old whiskey bottle that loggers used for kerosene to clean the pitch off their saws.

The forest on our land was first logged in about 1903. Horses or oxen pulled the logs out of the woods along skid roads down to the Cowlitz River. From the diameter and ring count of the old-growth stumps, we know the trees were all about the same age, around three hundred years old, and this indicates that a wildfire cleared the land around 1600. But there is so much we don't know. I want to learn the name of the man who placed the springboard on the stump next to Steve's Tree. Who was he working for? Where was he from? Finland? Germany? Or Minnesota or South Carolina? Or Vermont via Nebraska like the Willoughbys? Or was he a Cowlitz Indian?

What was the name of the man who greased the skid logs laid across the muddy road out of the forest? Was this his first job? Did he have a preference for what was in his grease pail? Dogfish oil? Animal fat? Rancid butter? What did he think of the timber strikes in Chehalis and Everett? What were the dreams of the man who shod the oxen? Did he favor one animal over the other? Did he worry that his job would be made redundant as more logging companies began to use steam donkeys to move the logs out of the woods?

A 1902 list of Toledo businesses includes two sawmills nearby: one owned by Benefiel Willard and another, Calvin and Son. In 1892, Calvin's mill produced the lumber to build the first bridge over the Cowlitz at Toledo. Perhaps the stumps with the springboard notches saw their trees go to one of these mills. Or perhaps the trees floated to Longview on the Cowlitz River and then down the coast to build San Diego and San Francisco. It is the kind of hyperlocal history that isn't often written down.

When I was teaching, I loved including primary sources and often asked students to use the Library of Congress's "Observe, Reflect, Question" method of analyzing a letter, political cartoon, or photograph. I like doing it now with artifacts from the history of our land. It frees me to ask any question, even ones that cannot be answered.

The two photos and one excerpt below are from *The Toledo Community Story: 1800–2019*, first published in the 1950s and periodically updated by the Toledo Historical Society.

Observe

Most of the men wear hats and suspenders, no coats.

Most men have some type of tool.

The men's pants are short.

They are standing on logs in the river.

It looks cloudy.

There are coniferous trees in the background.

The men look young.

They look like they have some kind of whistle around their necks.

The "unknown man" is partly hidden.

Reflect

It must have been fairly warm outside.

I wonder whether this photo is in the men's descendants' photo albums?

I wonder whether the men each received a copy of this photo?

The men must have had to work well together.

It seems such an ordinary workday, no celebration going on; maybe someone had a new camera to try out.

I want to know more about the unknown man.

Question

What was the occasion for the photograph?

When was it taken?

Where was it taken?

What were the different tools used for?

What is hanging around their necks?

What was the daily pay?

What were some "tricks of the trade"?

What made a good day on the job?

What part was more dangerous than another?

What kind of wood?

Where was it going?

For what purpose?

Who did they work for?

Observe	Reflect	Questions
None of the men wear hats.	Hats must have been disrespectful at a meal, even outside.	What is the food?
Many are wearing Civil War uniforms with two rows of buttons.	The Civil War uniforms must have been made well, or they wore them only on special occasions.	What are they drinking? Who made it? What is the occasion?
A place is set for each of them.		What is the man on the right holding?
All the serving dishes look like they're metal.	I wish I could smell the food and hear the conversation.	Why are some men wearing their uniforms?

This story was written by Joe Ryan, born in 1889, four miles east of Toledo on a homestead near Salmon Creek, a large creek that flows into the Cowlitz River.

From "The Ryans—A Beloved Family"

My father had one gear, and that was high and straight ahead.

He moved his family off the homestead about 1895 to a farm called the "Page Place," which was near his logging camp. For three years he held his logs for a better price in high water in Salmon

Creek to flush them out into the Cowlitz River. . . . The logs were contained in a big boom from which they were sold, rafted, and towed to the sawmills, which was the custom of many loggers along the Cowlitz River.

Then came the big flood, which rose to three feet of water on Third Street in Portland, Oregon. It carried all the loggers' logs at one time, breaking their booms. . . . The logs were lost by the loggers, only to be gathered up by the river pirates who owned tugs and could hide them in the sloughs and sell them later and pocket the money.

When my mother learned of this misfortune, she cried for days. All their hard earned money was gone. My father sold the bull teams and logging equipment and took a job driving ash logs down the Cowlitz River.

Observe	Reflect	Questions
Loggers sold the logs from the boom in the river, not at the mill.	His father's personality sounds like Dad's.	Who bought the logs? A representative from the mill? Or a middleman?
There were river pirates.	The flooded river filled with rushing logs must have been a devastating scene.	Who were the river pirates?
He had to go work for others after being his own boss.	It seems like he was logging trees from land he owned.	How many loggers worked at the logging camp?
Rivers were used to move logs.	Such a loss!	Did the father own the land around "his logging camp"?
The son describes the logging camp as "his."	There was no "social safety net." But there was still opportunity.	What were the ash logs used for?
		I wonder who lives on the "Page place" now.

At dinner recently, Dad mentions he can't find the kind of ax that was used to make the springboard notches. "Nobody uses them anymore."

Lou Jean, who has joined us for dinner, replies, "I think I might have one. I know I have lots of wedges, but I might also have that ax-head."

I feel like such a city slicker. I can't imagine having a collection of wedges, let alone a specialized ax-head, and I say, "I love you, Lou Jean."

She takes a sip of her wine and laughs.

Dad tells me he has some arrowheads he found on the tree farm in the
1970s. I ask him to show me. He opens the top drawer of his dresser
and starts naming what he sees:

> My grandfather's watch. That's my dad's watch there. They
> wore them on a fob in their vest pockets. These are all old coins.
> Another batch of coins—don't know the story on those—will have
> to go through them sometime. RHD belt buckle. Tape measures.
> Don't see anything right off. I'll have to do some more thinking.
> Scratching. A New Zealand fishhook. Oh, here are the arrowheads,
> in this box with some elk teeth. The elk teeth are ivory. People make
> necklaces out of these. Kind of pretty. It's like a present from the
> elk.

Dad says he found the arrowheads over on the west side of the road, past the spring. I ask why we haven't found any more and he says, "Nobody's looking."

The trees and soil that make up our tree farm are in the center of the traditional Cowlitz tribal lands. In the nineteenth century, the Cowlitz's understanding of land use did not depend on surveys, deeds, boundary lines, or individual ownership. My query of "who owned our land," first answered by a narrow ladder of names found in large ledgers, broadens out into an answer that changes the question, expands to thinking about an entire culture. I want to look, to know more about the community of people who were the first people to hunt deer and elk, gather camas and cedar bark, and drink fresh springwater from the land that now hosts our vegetable garden, empty doghouse, Gemini Grove, and the current clear-cut.

Many Northwest Native American tribes, including the Coast Salish, the larger linguistic group to which the Cowlitz belong, have a story about Raven. Steve loved this story, one he first found in Bill Reid and Robert Bringhurst's *The Raven Steals the Light*, and would tell it at moon-rising parties. As with all tales from oral traditions, versions abound and are shaped anew in each telling.

Steve stands next to the firepit at the bluff's edge. The moon is rising in a cloudless sky from behind Mount Rainier. He smiles and begins.

At the beginning of time, the earth was covered in darkness, an inky pitchy darkness. There were no stars and moon to light up the evening sky, and no sun to shine in the day. It was very difficult for anyone to hunt or fish or gather berries for food.

There was an old man who lived along the banks of a stream with his daughter. He had boxes within boxes and the smallest one contained all the light in the world. Every day he opened the boxes and said, "All the light in the world is mine, all mine. If no one can see my daughter, no one can know if my daughter is beautiful, and she will not leave me."

One day, mischievous Raven, who had always existed, overheard

the old man talking about his boxes. He instantly decided to steal the
light but first had to find a way to get inside the hut.

Lou Jean and I drive to BearRaven, home of Roy I. Wilson, honorary chief and spiritual leader of the Cowlitz. I've called to make an appointment to see the longhouse and museum he has built about five miles west of the Cowlitz River, and the tall, white-haired ninety-year-old is waiting for us on his riding lawn mower. As I've researched the Cowlitz people, Wilson's name appears everywhere. He has written over forty books and countless articles. I tell him I'm honored to meet him. He greets us warmly and asks where we began our lives: me, Ketchikan, Alaska; Lou Jean, Seoul, Korea. He tells us he was born on the Yakima Indian reservation.

He shows us a totem pole he carved that stands at the entrance to the meditation garden. The cedar totem is about twelve feet tall, and the eyes of many animals invite visitors in. He tells us the meaning of each creature. "The top is Bear, representing me, next is Raven, my wife's totem, next is the beaver, who controls the water of life. Under that are four carved faces: white, red, yellow, and black. No matter what your color, your race, your religion, you all are welcome here." Lou Jean and I nod, accepting his welcome. "The last carving is salmon because this is Cowlitz aboriginal land, and we are the salmon people. So Bear and Raven call on beaver powers to bring harmony and unity to all that come. My people, the Cowlitz people, welcome you to this site."

We sit in the longhouse and listen to him tell the stories of the totems holding up the roof. His storytelling animates him, and I raise my hand to ask the many questions from my research. I'm excited to have come to a main source. He laughs at my raised hand and tells me to interrupt him anytime. He tells me his Cowlitz name is Itswoot Wawa Hyiu, which means "Bear Who Talks Too Much." After an hour, I say we need to go, and I make an appointment to return the next week.

He focuses on Lou Jean. "You are from Korea?"

"Yes, I was three when I was adopted."

"Into a White family?"

"Yes, here in Toledo."

"Well, you are as American as all of us."

Lou Jean smiles and nods.

Back at the welcoming totem pole, Grandfather (as he likes to be called) shakes my hand. He takes Lou Jean's hand, kisses it, and brings her into a warm bear hug.

Steve shifts his body as smoke from the fire moves toward him.

The Raven watched the old man's house and noticed that each day his daughter went to the stream to fetch water. Raven transformed himself into a tiny hemlock needle and floated into the girl's bucket. Working a bit of his "trickster" magic, he made the girl thirsty and as she took a drink he slipped down her throat. Once down in her warm insides he changed again, this time into a small human being, and took a very long nap.

As I read about the first interactions between White trappers and the Cowlitz Tribe, I learn about Thas-e-muth, a daughter of the prominent Cowlitz chief Scanewa, and how she was given in marriage to a French Canadian, Simon Plamondon. The first White man to paddle up the Cowlitz River, Plamondon came to Scanewa's village, just south of what is now Toledo, in 1818. Scanewa's men seized him, and he was held captive for a time. But Plamondon could speak some Chinook, the trading language used among tribes, and soon Scanewa became fond of him. Scanewa adopted Plamondon and used him as a trader with Europeans in Astoria. As a wedding gift, Plamondon received twenty Cowlitz men.

Scanewa performed the marriage ceremony in accordance with the Cowlitz Sun Rites in a grove of trees on the north bank of the Cowlitz River. Sun Dancing was a time of celebration when Sun Dancers danced in an open field for three days. Most details about this dance have been lost; with the arrival of priests, traditional dances and rituals were discouraged and sometimes outlawed. Two of Thas-e-muth's younger sisters, who also married White traders, wed in Catholic ceremonies.

Thas-e-muth's place in history is a silhouette. She appears in articles and books about the Cowlitz Tribe, though her name is almost always replaced by "Scanewa's daughter" or "Plamondon's wife." I want to know more about her.

In a 1919 document naming the members of the Cowlitz Tribe, Thas-e-muth is named Veronica. Why was "Veronica" chosen for Thas-e-muth's Christian name? The traditions surrounding Saint Veronica resonate with Thas-e-muth's story. As with the young Cowlitz woman, Saint Veronica's life was bound by her connection to men. She used her veil to wipe Jesus's face clean as he carried his cross, and his face was imprinted upon the veil. Over time, this veil became part of many legends, including one in which it was used to cure the Roman emperor Tiberius of an illness. Sculptors have often carved Saint Veronica's image. In Sagrada Familia church in Barcelona, Saint Veronica kneels and holds her veil out to the world, clearly imprinted with the features of Christ. But the sculptor has made Saint Veronica herself faceless. I wish Thas-e-muth had access to the healing veil, for herself and for her people.

In 1838, after a Catholic mission was established near Toledo, a Father Blanchet baptized Simon and Thas-e-muth's children, Sophie, Simon Jr., Theresa, and Mary Anne. The entry in the record book refers to their mother as a "woman of the country, infidel."

Steve's dog, Zeppelin, comes to lie next to him. He reaches down to scratch behind the dog's ears.

> *The daughter went along with her daily chores not realizing that life was stirring within her. One day she gave birth to a child. If anyone could have seen him in the dark, they would have noticed that he was a peculiar-looking child with a long, beaklike nose, a few feathers here and there, and the unmistakably shining eyes of the Raven.*

Across the river from our tree farm, about a mile as the crow flies, the Saint Francis Mission still stands. A historical marker reads:

Cowlitz Mission

Venturesome frontiersmen, lured from civilization to the Oregon Country by the lucrative fur trade, so strongly besought leadership in religious worship for themselves and their families that Father Blanchet and Father Demers made the perilous journey and in 1838 founded here the Cowlitz Mission. This oldest mission in the northwest now stands where then, in rude dwellings, the beneficent fathers held service for the pioneers and with simple picture writings, not unlike their own, taught the Indians religious history and the blessings of devotion to the Great Spirit.

I visit the cemetery next to the brick church. Simon Plamondon, who died in 1881 at ninety-nine years of age, lies buried next to a row of hundred-year-old cedar, but Veronica is not there. Perhaps upon her death she was placed in a ceremonial canoe in the Cowlitz tradition: face down, her head pointing to the mouth of the river, shells on her eyes, and blankets and mats covering her body. I do find the grave of one of her granddaughters. The inscription reads: "Mary Louise Theresa Plamondon Hays Bouchard Wilson King, September 30, 1875– September 7, 1957. Granddaughter of Pioneer Simon Plamondon and Veronica, Daughter of Principal Chief Schanewah Great Cowlitz Nation Matriarch, Long Line of Cowlitz Sons and Daughters."

The Cowlitz Tribe now owns buildings adjacent to the cemetery that were once Saint Mary's Academy, a girls' high school that operated between 1911 and 1973. The tribe uses the site for meetings and has created housing units for some of its 4,100 enrolled members. In the late 1970s, for a brief four years that corresponded with my high school years, Cowlitz Prairie Baptist Church owned these buildings. I was an active, fervent member of this church and spent countless hours there but knew only of its recent, Catholic history. I knew that the Baptist leaders had taken out the suspicious confessionals and kneelers, but my knowledge of the Cowlitz people was vague and formless.

After visiting the mission cemetery, I walk over to the chapel to see whether I can get in. It's open, and I stand in the Catholic-

Baptist-Cowlitz sanctuary and reminisce. I hear the voice of my youth pastor singing, "There's a sweet, sweet spirit in this place." Down the hallway is a gym with a stage where friends and I performed the musical *Show Me Jesus*. I think of the basement with foosball tables next to the room we used for Wednesday Bible study. I have a memory of sitting on my boyfriend's lap as the youth pastor exhorted the gathered teenagers about the virtues of abstinence. During my junior and senior years, the church hosted a few families of Cambodian refugees, and the aromas of their cooking come back to me as I walk the halls.

I'm glad the Cowlitz have this land back. The land supporting these buildings was not part of the land grant process—it was given directly to the Catholic Church by the Cowlitz Tribe in the 1800s. Many tribal members feel that the Catholic Church should have given the land back rather than asking the tribe to buy it. A magnificent bigleaf maple stands in the lawn next to the housing units, and I wonder whether it is old enough to have witnessed the transfers over the past two hundred years. I hope the trees and streams continue to outlive the vicissitudes of human activity.

Dad adds a log to the fire. Sparks fly and pitch crackles. Steve resumes the story:

> *The Raven boy loved to talk and sing, but only the daughter could understand him. "What is he saying, daughter?" "He wants to play with your boxes." The old man could see no harm giving the boy his biggest box, and that satisfied Raven for a while. But soon he wanted the next box and then the next, until finally only one box, the box that held the light, remained.*

On my second visit to Roy Wilson, he suggests that I go see Michael Hubbs, a genealogist for the tribe, and I make an appointment. His office is in the Cowlitz tribal buildings I've just visited. Chart paper with Lower Cowlitz vocabulary lines the walls and windows of Michael's office—he teaches language classes on Saturdays. Hubbs tells me about a few of the sources he has used in his genealogy work:

journals kept by two Hudson's Bay traders at Fort Langley, British Columbia, twenty-five miles east of present-day Vancouver, and Fort Nisqually, eighteen miles south of present-day Tacoma, Washington, as well as records kept by the first Catholic missionaries in Toledo. These journals change the way I've been thinking about Thas-e-muth. I'd imagined her living three miles down the Cowlitz River in Tawamiluhawihl or Matup, riverside villages of around three thousand people. Thirty-eight cedar longhouses once stood there, built perhaps of trees felled on our future land. Six-fire longhouses sheltered twelve families; nine-fire longhouses, eighteen families; ten-fire longhouses, twenty families. I'd imagined Thas-e-muth among these families, gathering sword ferns to line a cooking pit for salmon; giving her children chewing gum made from fir sap; weaving cedar-root baskets; and making cedar-bark clothes.

But the journals reveal that for at least a few of her short years she lived within the walls of Hudson's Bay Company forts. What did she do while Plamondon was laying foundations for HBC buildings, trapping otter and beaver, squaring rafters, hunting deer and elk, or fishing sturgeon? The journals indicate that in 1827 her father, Scanewa, and his family also joined an HBC expedition from the Cowlitz River to Fort Langley. Perhaps Thas-e-muth spent her Langley days with her father and siblings? In 1829, she gave birth to a child there.

The Langley trader's March 12 entry for that year reads: "At a few minutes past midnight, a Girl was born to Plemendon, at precisely 2 a.m. our party was under weigh [way]—They have near upon 100 Skins in Trading Goods, in Case the Indians of the Sound may Come to them Enpassant with Beaver."

In 1830, Thas-e-muth is with Plamondon at Fort Nisqually, where he is busy building cupboards, wooden scales, beams, a counter for the Indian Shop.

The trader's July 11 entry reads: "All last night the Indians nigh us were singing to a medicine man who was doing his best in the killing of Plomondon's wife who has been sick for some time, I have endeavored to stop the business but believe to no purpose as she is bent on getting blowed by her country-man. Fair weather."

The trader's July 20 entry reads: "Plomondons wife has been unwell some time, and all her care is to give away property to Indian Doctors for curing her, though at times she applys to me for medicines, which are given, but the relief she gets is attributed to her Doctors. Fair weather."

The journal keeper makes no note of Thas-e-muth's death, and later documents do not agree on the date, variously indicating that she died in 1828 or 1830.

While still in his office, surrounded by Cowlitz words and their English translations, I ask Michael about his genealogy. He tells me that he is a direct descendant of Simon and Thas-e-muth, through their son Simon Jr. I am awed. I ask whether I can touch him, and embarrassment strikes me immediately. My curiosity has been momentarily crystallized into a fixation that idealizes and separates. It has turned Michael into a curiosity; he is uncomfortable too, but gracious. He tells me more about Simon Plamondon Sr.; he is not buried under his headstone, but in another corner of the cemetery. Augustus Bonnin, one of Simon's pallbearers, told a Mr. Borte, who told Michael. The name Bonnin sounds familiar, and back home my research confirms that Augustus Bonnin and his wife sold the land that is now our tree farm to the Boones from whom we bought!

I ask about Thas-e-muth's burial, and Michael says that based on his research, she had a Cowlitz burial whereby her bones would have been buried about a year after her death. He thinks her bones are near the site of Simon's cabin—between present-day Toledo and the mission, across the river from our tree farm. Michael and some others have found two mounds covered in underbrush but have not made the location public.

Michael also tells me that while northern tribes such as the Makah, Tlingit, and Haida carved tall totem poles featuring many figures human and animal, the Cowlitz and other Salish tribes favored single-figure carvings as signs of welcome to their villages. I go to see one at Evergreen State College. A bright drummer painted red, green, and white stands straight and tall and I imagine the sounds of the drum spreading out over the verdant campus.

I wonder what Thas-e-muth thought of Simon, of the Hudson's Bay Company, of her children's prospects. We have none of her words, yet her story lives on. She is an old-growth cedar stump where young hemlock and red huckleberry take root.

Steve takes a sip of his beer.

The Raven boy asked for the smallest box and saw the light inside. After much coaxing and wailing the old man at last agreed to let the child play with the light for only a moment. As he tossed the ball of light the child transformed himself back into the Raven. He snatched the light in his beak, flew up through the smoke hole, and soared into the sky. The world instantly changed forever. Mountains sprang into the bright sky and reflections danced on the rivers and oceans.

Thas-e-muth died at the beginning of a flu epidemic that wiped out around 90 percent of her people. It's thought that the epidemic of 1829–1830, called the "gray" fever, was brought in by Captain John Dominis, who traded blankets and guns for furs from his ship, the *Owyhee*. Claiming the Cowlitz were saving the best pelts for the British, Dominis called them together and held up a vial. Though it looked empty, he told them there was sickness inside it, and he threatened to release it among them if they didn't bring him the finest of the beaver pelts. The Cowlitz refused and Dominis took his vial and crew, some of whom were ill, leaving behind exactly what he had threatened—a disease the Indians came to call "cold sick."

The botanist David Douglas describes that year's epidemic in a letter to William Hooker:

> A dreadfully fatal intermittent fever broke out in the lower parts of this river about eleven weeks ago, which has depopulated the country. Villages, which had afforded from one to two hundred effective warriors, are totally gone. Not a soul remains. The houses are empty and flocks of famished dogs are howling about, while the dead bodies lie strewn in every direction on the sands of the river.

According to Hudson's Bay Company records, in 1828 there were fifty thousand Cowlitz Indians. By the time the Willoughbys arrived in 1884, only a few scattered communities remained.

The moon is rising over the horizon now. It is so clear we can see it reflecting on Mount Rainier's snowy top. We pause to exclaim, then Steve's voice resumes:

> *The old man transformed into an Eagle and began to chase Raven. Raven was so caught up in all the excitement of the newly revealed world that he nearly didn't see Eagle bearing down on him. He swerved sharply to escape the Eagle's outstretched talons, and the ball of light crashed into the side of a mountain, breaking off into one large piece and many small ones. The bits of light bounced back up into heaven, and they remain there to this day as the moon and the stars.*

I decide to go to the 2018 Cowlitz PowWow at the Clark County Convention Center. Under HVAC ducts and fluorescent lights, folding tables create a walkway inside the cavernous hall. Each table announces a different organization: Cowlitz Department of Public Safety; ilani Casino; Division of Child Support; Cowlitz Tribe Health and Human Services; Cowlitz Tribe Treatment Center. Vendors and their tables continue to shape a path along the cement floor: drums, clothing, fleece blankets, backpacks, jewelry, candles. One woman is selling cards with messages like

> You say a bunch of immigrants are refusing to assimilate to your culture and threatening your way of life? Man that sucks.

> Hate to tell you this, but you're all illegal aliens.

> Regard Heaven as your Father, Earth as your Mother and all That Lives as your Brother and Sister.

I buy one that pictures an elderly Indian woman in front of a weathered wooden fence. It reads: "Consider the advice of your elders: Not because they are always right, but because of the wisdom they have gleaned from being wrong."

The longest line is for food: fry bread, Indian tacos, nachos, pumpkin pie. I've just finished listening to Tommy Orange's novel *There There*, and images come to me of young cousins gathering their money for Indian tacos at the Big Oakland PowWow, and of a white 3D gun and a sock full of bullets thrown over a wall. Listening to the book as I drove back and forth to the tree farm, with Orange's chapters building to an unflinching bloody end, I had to take a break. For two trips up and back, I chose to listen to silence for an hour and fifteen minutes, unable to face the trauma he'd set up. Now my stomach constricts and I try to put the violence of that fictional PowWow out of my mind.

The program says the Grand Entry is at one o'clock, so I find a seat on one of the metal bleachers set up in a semicircle on the cement floor. I don't really know what a Grand Entry is, but people of all ages are gathering at one end of the hall, dressed in beaded and feathered clothing. The Host Drum, *4 Bands* from White Swan, Washington, sits in a circle next to the emcee. Veterans carry flags representing the US Army, Navy, and Marines, as well as the US flag, the Cowlitz Nation flag, and the POW/MIA flag. One speaker thanks the veterans, saying, "Without them, we wouldn't be here on our aboriginal land." Roy Wilson gives the opening prayer in Chinook and English: "To the Chief of the Above, we are the Cowlitz People and friends. We have come to this place to have a good time. We ask you to give us good singing and dancing."

The drummers begin. Five men hit a large drum, softly then loudly, and sing in a haunting high cry. Dancers circle the floor, some walking, some doing a double step, some twirling arms and lifting their knees high. The drums and singing transport me. The bright lights dim and the cold hard floor softens. Time feels fluid and the past, present, and future flow together. A sound from my history reverberates: noises of Obon dancing, festivals in Japan with drumming just on the downbeat, no syncopation. The drums finish and my metal seat starts to feel cold and uncomfortable.

On the way out, I notice a man setting up a table display of art that intrigues me. On old maps and on ledger pages, much like the pages of books recording our land's ownership history, he's made bright paintings of canoes, birds, warriors. He hands me his biography on a laminated card. I learn that he is Robert "Running Fisher" Upham and that his work is "Ledger Style Art," in the tradition of the Winter Counts of the Plains Indians. Imprisoned Plains Indians, separated from animal skins, drew on discarded accounting ledgers. The spaces of the ledger become part of the art, merge with the new drawing, and create a counterpoint.

An analogy comes to me: a ledger is to painted birds and warriors as concrete and HVAC ducts are to dancing and drums.

Steve chokes up as he begins the ending of the story:

Eagle pursued Raven beyond the rim of the world. Exhausted by the long chase, Raven let go of what light still remained. It floated gracefully above the clouds and remains there to this day as the sun. The next day, the first rays of the morning sun brought light through the smoke hole of the old man's house. He was weeping in sorrow over his great loss. Then he looked up and saw his daughter for the first time. She was very beautiful and smiling, and he began to feel a little better.

PART FOUR

Sapwood

Sapwood is the tree's pipeline for water moving up to the leaves. Sapwood is new wood. As newer rings of sapwood are laid down, inner cells lose their vitality and turn to heartwood.

Massive steel ships carry logs grown in our soil across wild salt water.
I want to join our wood, the flesh born from the Cowlitz, Willoughby, Boone, Stinson soil, as it travels the oceans to its new homes. I
imagine myself in a cabin on the ship, looking out at gray, absorbing
the contrast of elements and industry. I will use the solitude to write,
to shape the notes I've been gathering about my family and the land.
And I can gather new notes. A ship's voyage across the Pacific tantaliz-
es with its visual and psychological unknowns.

I start looking for a way to get aboard a log ship. A family friend,
Mike Warjone, at Port Blakely Tree Farms says he can help. He'll con-
tact Nippon Yusen Kaisha, NYK, the company that owns the ships
Port Blakely charters to export wood to Asia.

I receive an email from Mike:

It looks like the cost of the voyage will be minimal. ($14/day)
The disembarking paperwork in Japan will need to be sorted out.
The ship will likely have a Japanese captain and a Filipino crew.
You may be required to do some chores while on board.

The shipping company has apprehensions about sending a woman to sea with a dozen Filipino sailors and no particular rules of engagement in place. We can talk about the perceived vs real safety issue. I suspect that if you are on board as a journalist, you will be fine.

The food does taste like crap. Lots of weird bonefish soaked in oil purchased in bulk from Hong Kong.

Once you have a timeframe in mind, we can circle back to John and the shipping company.

I start telling friends about my adventure. Most are skeptical, but I am all anticipation. The upcoming voyage becomes part of the story I tell—at cafés and bars, on walks in Forest Park, and at the tree farm table during dinner. But then, I hear nothing for a few months.

One morning, as I'm scouting for honeysuckle, my cell phone rings in my shirt pocket. It's Mike. I sit on a stump to talk.

"NYK is really nervous about you going alone. Do you have a friend who could go? I'd go, but my wife says if I'm going to spend that much time away from our daughters, it has to be with her."

"The only person I can think of who would want to be on a log ship for twenty days is Dad. I'll ask him."

I don't want to ask him. I want to be away and alone. But it looks like I can't go unless I do.

Never, until now, have I wanted to be a man.

Dad agrees immediately, and I'm glad he is happy. But the nature of the trip has changed. I'll be able to feel space and time recede, and I'll be able to talk with the sailors, but I will not be away and alone. Much will be asked of me. Dad is a big presence.

Mike tells us we have a meeting with Captain Sakai from NYK. I start writing questions:

What port in Japan?
Is there Wi-Fi?
Will I be able to do laundry?
Will there be a refrigerator and microwave I can use?
What types of jobs can I help with?

Dad starts telling me questions I need to ask. His questions are already on my list. I sigh, "Dad, we need to take a field trip to the shop." He frowns but gets up from the computer and follows me across the driveway. We walk past the chainsaws, the spray tanks, around the table saw to the sign above the tool bench, and I ask him to read out loud: "Do Somethin' Either Lead Follow or Get the Hell Out of the Way." He looks confused. "Dad, I am the leader on this trip. You need to follow." His face clears with understanding, and he bows to me.

And he continues to tell me questions I need to ask. I leave my list on the dining room table so we can both add to it as we think of questions. He reads the list. "That's good. I can't think of anything else." I hear him on the phone telling his brother, "This is Ann's trip, part of the book she is writing." He's trying.

I want to go to the meeting with Captain Sakai by myself, but Dad says we should go together and "show them 'our' power. We are a powerful duo." How can I maintain my power next to his? I want to ride the log ship with Dad as a companion and fellow traveler, not with my guard up protecting myself and my story. He will be part of my story. He already is. How can I be the main character of this chapter, not just an observer of Dad and the others?

Dad and I drive past decks of logs to Port Blakely's office at the Port of Longview. Mike pulls in, unfolds his six-foot-four frame from his pickup, and walks up to us. "I don't know what this meeting is about," he says with a frown.

In the meeting room decorated with diagrams of Japanese houses made with Pacific Northwest wood and photographs of log ships, Captain Sakai tries to dissuade me:

"It's so boring."

I want to be bored.

"There are no doctors."

I am healthy.

"The food is terrible."

I don't care.

Mike senses trouble. "The Stinsons have been family friends and business partners with Port Blakely for decades. They are pillars of the local timber industry. Please do whatever you can do to make this trip possible."

Captain Sakai looks miserable. It is clear he's been given the task of discouraging me.

I try to build some rapport. He has a Seahawks sticker on his laptop. "Are you a football fan?"

He nods. "Yes, but baseball is my sport. I've taken my four-year-old son to some Mariners games."

The atmosphere still feels cool; this sports chat is not working. I try another approach.

"Can I go?"

"Such a direct question. It's not my decision . . . my bosses are concerned. They suggest a short trip either around the Olympic Peninsula from Olympia to Grays Harbor or between two ports in Japan."

It's a no without a no.

Outside, we stand next to Mike's pickup. Tears try to spill out. Mike says, "So that's what the meeting was about. But I'm not giving up. We'll get you on a ship even if it's just for a short trip."

On the way home, Dad drives and I lean my head against the window, a year of anticipation coursing through me with no outlet. Where do I put all my yearning? And how did my yearning get so big?

Mike makes good on his promise. He puts us in touch with Ken Shioda, a marketing representative for Port Blakely. Ken is a gregarious man with an easy laugh. As I share my desire to be on the ship, his first quizzical face relaxes into understanding. "You want to know who touches the logs!" he says exuberantly. Yes! He tells us that we will be able to board a log ship in Japan as it travels between two ports. He has also arranged for us to visit some mills that process trees from the Pacific Northwest. And he will be our guide.

My heart starts to warm to the shorter journey in Japan. MV *NEW DAYS*, an NYK vessel, will be loaded at the Port of Longview between March 11 and 14. It takes about a week for longshoremen to fill the ship with logs. We can tour the ship, meet the captain, and then reconvene in Japan, joining the crew on the twenty-four-hour journey from Nanao to Maizuru, two ports on the Japan Sea.

As I prepare to visit the ship in Longview, I have an idea. I take Instax photographs of Mom, Dad, Lou Jean, Izzy, Tom, and me. These I tape to the first pages of a notebook and write notes about who and where we are. I put the notebook, camera, tape, and pen in a box. I will ask the captain of MV *NEW DAYS* to take photos each day of the voyage, tape them in the notebook, and write notes. I want a physical record, a ship's log. The camera replaces my eyes on the trip. The notebook may not be sentient or able to feel the days go by, but perhaps it will capture some sense of time passing. A few days later, when I bring this idea to Captain Sakai, he pauses for a bit, then agrees, his shoulders relaxing, a small smile on his face. I think he is happy to be able to agree to my request.

I want to know about the history of ship logs and I learn this:

> The origin of the logbook, and of its basis as a tool of navigation, mediating sailors' knowledge of their movement through space, lies in that most material of artifacts, a wooden log. In his 1574 *A Regiment for Sea*, William Bourne describes how, with a line tied to it, a log was thrown over the side of the ship, and the length of line was measured "while someone turned a minute sand glass" or in some cases "spoke some number of woordes [*sic*]" in order to measure the ship's speed. ("Logging-In: The Ship's Log as Medium," Janet Sorenson)

At the dock, a sailor takes my bags. I grab a rope banister with my left hand and take his free hand with my other. The step down is half the length of my legs. MV *NEW DAYS* looms above and beside me, its 24,655-ton mass stretching beyond my peripheral vision.

Inside the ship, we climb three flights of stairs. Light green paint brightens the steel walls. Sun rays reflect off the Columbia River. The captain of MV *NEW DAYS*, Alfonso Constantino, comes into the office; he's a soft-spoken man of about fifty. He puts on his glasses as I show him the journal's sample pages (I feel like such a schoolteacher) and explain how to use the camera. He smiles and agrees to my project.

As we leave the ship, we pause on deck and watch men operate loaders with huge claws swinging entire loads of logs from trucks to the deck of the ship. Some logs are dark with bark, others light after debarking, and the contrast creates a pleasing design. The shape of each still resembles the tree from which it came. The forest is not far away. But as the logs cross the ocean, daily moving farther from the soil that fed their roots, they inch closer to the saws, dryers, and conveyor belts that will transform them to wood, move them out of industrial spaces to domestic areas. I want each builder, family, or congregant to look at the wood and see when it was a log and feel when it was a tree.

As MV NEW DAYS travels across the Pacific, I take a journey of an inch to Reichert Shake and Fencing Mill just a mile and a half from the tree farm. We sell them all our cedar logs, and their crew of sixty saws the logs into fence boards.

I sit in the office of Russell, mill supervisor, down the hall from the visitor waiting area with its mounted heads of Dall sheep, Roosevelt elk, and white-tailed deer. I've brought a plate of cookies, apple spice and chocolate crinkle. I share Steve's belief that baked goods from one's own hands open mouths and hearts. My first appointment with Russell had been canceled when too many employees failed to show up for work, and he was obligated to work on the line. I ask about keeping workers. Russell sighs and says roughly half his people have been with the company for over ten years, a few for thirty. But seeing new employees through their first year is always a battle. Sixteen hires from this year have stayed, but thirty-seven have come and gone. The drug test challenges many.

I ask about the demographics of the crew. Thirteen are Hispanic, the rest White. Russell says, "We have everyone here. Three or four religions. I even have an atheist. He works next to a Jehovah's Witness. They were getting into it, so I brought them into the office where they could have it out. They're friendly now."

I follow Russell through the carpeted main office, across gravel, over to the cement-floored mill. He takes me first to the "shack" of the sawyer, Dale, who has been with Reichert's for twenty-three years. His job is to make the first cuts on each log that comes into the mill. From an office chair with a flowered cushion on the seat, he looks through plexiglass and adjusts red laser beams that tell the saw where to cut. I ask what makes a good day at work. Dale says, "A steady supply of wood, a good mixture of sizes, the machine not blowing out lines."

At 10:30, Russell takes me to the lunchroom, where about ten guys are eating. He sets the plate of cookies next to the row of microwaves. Hands reach and the pile is halved in seconds. I ask what makes a good day at work.

"everyone's actually here"

"we have a full crew"

"everyone's doing their job"

"nothing's broke down"

I ask what they do to help new workers show up.

"talk to them"

"pick them up if they need a ride"

One kid, Tony, wiry and jumpy in his jeans and cotton work shirt, says shyly, "They gave me oil for my car motor the other day. I wouldn't have been able to get home without the help." He heats up leftover spaghetti and settles in a folding chair to eat.

Ten minutes later, the horn blows and the lunchroom empties. I take photographs of the strand-board walls. Color printer photos of deer and elk, OSHA posters, a whiteboard with handwritten ads:

1985 Honda 200X ATC, completely restored $1,500, Call Joe 360 xxx-xxxx. Chiquita, Dole, and Del Monte stickers decorate the bottom left of the whiteboard.

Russell picks up what's left of my cookie plate, and we head to an office in the mill. One by one he calls workers not needed on the line to come in and talk with me.

Crystal. Forklift driver. Eleven years.

The first year, she worked nights, 3:30 to midnight on cleanup, then four or five on the green chain, pulling freshly sawn boards off a conveyor belt. She says she'd like to retire from here. She's proud they grade their own wood rather than relying on computers to do it.

Derek. Millwright. Nine years.

He was heading down the wrong road. Wanted to quit high school and work at the mill with his dad. Russell, Derek's dad, and Reichert's owner made a deal that Derek could work at the mill from 3:30 to midnight as long as he stayed in school and kept his grades up. In 2009 he graduated from Centralia High School and has since worked his way up to millwright. He's had a few offers but won't leave. "It's family, close to home. I like being with a crew that gets along." Derek's house is now home to a new employee, "a guy having a hard time, was living in his car." He also helps train new guys. He says it used to take two weeks to see whether a new guy would make it, but now it takes a month because the first two weeks are busy with basic skills training: how to use a shovel, a broom, a rake.

Randy. Head electrician. Twenty-one years.

Twenty different tools line the front pocket of Randy's overalls. I ask how the tools get there each day. "I place them on my toolbox at night and put them in my pocket each morning." I ask him to talk about three of them. "An AC tester—used for quick troubleshooting, called a 'wiggy.' A voltmeter—it measures volts, amps, and ohms and fits in my pocket. And a small screwdriver—can use it for everything—cleaning, popping out O-rings, stirring coffee."

Verses from "Frozen Logger," a song we sing with Mom, come to mind:

As I sat down one evening, within a small café,
A forty year-old waitress, to me these words did say:

"I see that you're a logger, and not just a common bum,
'Cause nobody but a logger stirs his coffee with his thumb.

My lover was a logger, there's none like him today.
If you'd pour whiskey on it, he would eat a bale of hay."

Derek and Randy sit on two folding chairs. I sit on an office chair missing its upper half and Russell sits at a desk with a computer screen that shows him what's happening around the mill. A black plastic garbage can rests between Derek and Randy, and they use it as their spittoon.

Randy loves the local area. He says, "I don't like going north of the river [two miles from the mill]; there are city slickers there." This reminds me of a college student I met in New York City who said he didn't want to live north of Fourteenth Street after graduating.

Randy and Derek tell me the mill feels like a family. "It's hectic. There are ten things that need to be done, ten different ideas on how to do each, and we have to come to an agreement. And we do."

All three walkies buzz. The chipper's plugged. Randy and Derek groan and get up to leave.

I want to know more about Tony. Russell tells me he almost fired him a week ago. Tony was always late, absent, sulking around when at work. Then one Monday he came in ready to work. I ask Russell to show me where Tony is working on the line. We walk to the four- and five-foot green chains. Workers pull boards from a conveyor belt and stack them on carts. Tony is hustling, almost running from the belt to the cart in his gloves and safety glasses. Russell calls him over, but Tony indicates he can't leave his position. His eyes are wide, wary. Russell says, "You're not in trouble," and Tony walks over. Russell asks, "What changed, what helped you turn a corner?" Tony replies quickly, "I wanted to get my self-respect back. I wanted to be part of a team," and he heads back to work.

The few hours with the people at Reichert's cause a small joy to well up; our wood helps keep this company in business, helps keep the people here part of a team, a family almost.

In February 2020 as I work to revise this book, Dad hears that Reichert's has been sold to a competitor and will shut its doors. The news travels quickly and shocks our tree farmer friends, but the local papers do not report on the closing. I find some brief details on the website of the company that advised Reichert's during the sale:

> Reichert was acquired by Alta Forest Products, the world's largest producer of wooden fence boards, specializing in Western Red Cedar and other high-grade species. Alta Forest Products is owned by the Itochu Corporation of Japan. Itochu Corp. owns several other fence manufacturers and distributors, including Master Halco and Jamieson Manufacturing Corporation. The acquisition of Reichert enables Alta Forest Products to expand its production capacity and strengthen its position in the market.

I drive over to the mill site. Across Kangas Road from the empty buildings, fluted cedar butts mark the boundary of the gravel parking lot. I shut off my engine, wondering whether I am near the spot Tony received oil for his car. I chide myself for feeling so nostalgic. Family companies get bought out all the time. But I can't help worrying about the employees and their loved ones.

I look through my windshield. "No Smoking" and "Employees Only" signs hang on wooden posts, but the sheet-metal buildings are silent. Muddy cedar bark lies piled up behind the guard house. The log yard has no logs. The lunchroom has no people. The saw shop has no saws. No conveyor belts or screwdrivers used to stir coffee. No adults gathering to help a young man find his way.

Rain begins to obscure my view.

I say goodbye.

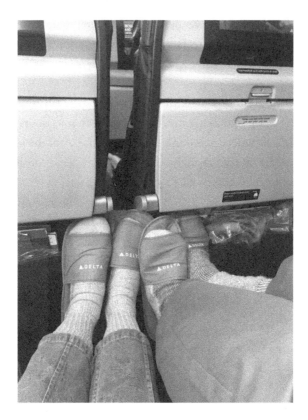

Dad and I fly to Japan. Seats 35A and 35B. No wind on our faces. No rolling ocean waves. Nine hours race us through the longitudes MV *NEW DAYS* takes twenty days to traverse.

At Narita International Airport, we buy two cans of beer, eat a bowl of fast-food ramen and a cellophane-wrapped rice ball, and wait for our second plane. It makes me smile to pay with yen for the first time in two decades.

In college, I chose to study Japanese, partly because Japan had always seemed close to me; the near west rather than the Far East. When we were young, Dad pointed out trees that would make "J-logs," logs to be exported to Japan. Gifts from Japanese companies often arrived on holidays; a favorite one for us children was an ice bucket, its lid decorated with ceramic olives. At our parents' parties, we'd put real olives on the lid, pass it to guests, and laugh hysterically when they tried to pick up a fake one.

This trip merges deep interests: Japan, forests, a curiosity about the history of things, and curiosity about people. For fifteen years after college, Japan was a central part of my life: two years teaching English, one in Fukui, one in Chiba, where I lived with a local family; an eight-year marriage to a Japanese man; hours and hours of kanji study, a master's degree in Japanese history. In recent years, I've wanted to go back to Japan, but not as a tourist; following our wood here has given me a reason. Dad and I will also go to Korea, a country neither of us has visited, where we want to see the temples built with Pacific Northwest wood.

The flight to Hiroshima is gorgeous, skirting Mount Fuji as the sun sets; a pink sky outlines a silhouette of its iconic cone. Dad climbed this mountain five times in 1957 while stationed at a nearby Marine Corps base, and he smiles to see it in its glory.

On our first morning in Japan, Dad and I wake early and walk outside to see what we can see. The sky is blue in Kure, a port city surrounded by green mountains. We find a river lined with cherry blossom trees in full bloom and follow it toward the sea.

Ken has arranged several tours: sawmills owned by Chugoku Lumber Company and a house being constructed with the lumber they produce. The Chugoku Timber mill in Kure is the largest Douglas fir mill in Japan. Each week, a ship carrying five million board feet of logs comes from Longview direct to Kure. I have heard about this mill from Mike Warjone and others who have visited and been astounded at its size and efficiency. I'm excited to see it for myself.

Because flat land is scarce, the mill is four stories high. Raw logs start their journey on the first floor and products are finished on the fourth. At one metal landing, our guide, Taka, shows us a view of the Seto Inland Sea and green mountains beyond stacks of logs and banded lumber on concrete.

We walk through the lunchroom. Rice steams in the rice cooker next to boiled eggs, tea, soy sauce bottles, and chopsticks. From the kitchen, the salty smell of miso soup and pork cutlets entices me. Taka informs us the company pays half, so lunch costs just two dollars.

At ground level we walk between the rows of logs. They stand thirty feet high and create a narrow corridor. Dad touches the round ends. "Beautiful logs. Look at the ring count. See how small the knot size is? No sweep, tight grain, all the best logs from our land end up here."

Taka drives us to the next mill. As we walk in, we pass stacks of wood labeled "Miyazaki home, 3rd floor," "Tanaka home, 2nd floor": the lumber here is cut to specifications for individual houses. Machines glue smaller pieces of wood into large beams, strong Douglas fir on the outside, lightweight Japanese cedar on the inside, making sturdy, light supports. Workers in another area operate machines that cut notches in the lumber. The end product begins to resemble Lego parts. The mill manufactures enough lumber for about seven hundred houses a month.

In cavernous concrete and steel buildings, I watch the logs being sliced and diced, rolled along conveyor belts, stacked, tested for strength and moisture, dipped in an antibacterial solution, sanded, and glued. The whine of saws and the sucking sound of hydraulic pumps hurts my ears. The sharp metal flaying the flesh of the trees hurts my heart. Concrete covers the earth. Green steel blocks the sky. Industrial fans blow hot air and suck sawdust. Workers operate machinery, rarely touching the wood.

I can find beauty in logging, in slash burning, in stumps, but this leaves me cold. I want to go back to the days of hand-hewn boards. But then I feel silly, like a disconnected urbanite wanting "small-batch," "artisanal," "single-sourced" products that by their nonindustrial origin are available only to people with wealth.

A more trained eye would see the technical expertise and be in awe of the feats of mechanical engineering. But I don't know enough to see. In 1981, at Toledo High School, all seniors took the Armed Services Vocational Aptitude Test. I earned two out of twenty on the mechanical reasoning section. And I haven't found room in my life to become curious and learn. I regret this now.

I wonder how Steve felt about screeching sawmills. I know he wanted to find niche markets that would allow us to grow our trees longer than the industry standard. He wanted the forest to have another twenty or thirty years to mature and age, creating different wildlife habitat, cooling streams, and sucking carbon dioxide from the air. But I wonder whether his desire was also so our trees would have a less industrial end? I wish I could ask.

We pull into a neighborhood street and stop in front of a two-story house being built with Chugoku lumber. It's only the fourth day of construction, but the entire frame is in place. Because the wood has been precut at the mill, two carpenters can put up a house in about three months. I want to know whether any of the wood will be visible when the house is complete. The answer is no—Sheetrock will cover the inside and brick or tile the exterior. I feel unreasonably disappointed. Why do I want all our wood to be made into a thing of beauty? Why isn't sustainable, affordable, and strong good enough? At least the house isn't being built with concrete and steel. Steel production begins with extracting iron ore from the ground, never to be replaced. Wood "production" means growing trees, sustaining forests. I need to make peace with the fact that our trees may have a prosaic rather than an artistic end.

As we drive back to the hotel, I tell Dad that we have the best job in the timber industry. Working in our forest, tending the trees as they grow is the best place to be. He smiles. "That's why I left the mills after twenty years. As a tree farmer, I'm living my dream. Just being out and placing a piece of slash next to a newly planted cedar, providing it shade for the summer. It doesn't get better than that."

In Nanao, a sailor takes our bags. The morning snow has melted and the sun is warm on my face as I walk up the metal steps onto the deck of MV *NEW DAYS.*

The sailor guides us up the same metal stairs we climbed in Longview, this time one more floor up to the bridge, the room that houses the steering and communication equipment and where we will spend most of the next twenty-four hours. The view is spectacular: the bridge sits high above the deck of logs and I can look out in three directions at the Sea of Japan and its islands.

A harbor pilot has come aboard to help guide the ship into the sea. As he begins giving directions, the immense ship starts to move, the motion almost imperceptible. The instructions sound like a liturgy, a call and response; the pilot's command, a repetition from the third officer, then a confirmation the command has been completed.

Pilot: *Starboard 10*
Third Officer: *Starboard 10*
Third Officer: *Starboard 10, Sir*
Pilot: *Slow Ahead*
Third Officer: *Slow Ahead*
Third Officer: *Slow Ahead, Sir*
Pilot: *Starboard 15*
Third Officer: *Starboard 15*
Third Officer: *Starboard 15, Sir*
Pilot: *Full Ahead*
Third Officer: *Full Ahead*
Third Officer: *Full Ahead, Sir*

The pilot bows, his job done. He leaves the bridge, and I see him appear next to the log decks four floors below. He climbs over a railing, down a long rope ladder, and into a tugboat that has drawn up next to MV *NEW DAYS*. The ship is now free to sail to Maizuru.

I step outside into the air. Wind blows at my hair. I am on the ship! I am with the wood on the water. I move around a corner so no one can see me, and I give a loud whoop.

Captain Alfonso comes to greet us. In his hand is the ship's log box, and he's done it! I start to hug him but choose a handshake instead. My notebook, my insentient substitute, has been returned. The captain's photos are taped to each page and his neat penmanship provides more details. Distant experiences are in my hand.

I open the ship's log and study the crew's photos from the twenty-day voyage.

Captain Alfonso is in charge of a crew of nineteen. All are Filipino, working a nine-month

12 March 2019 Capt. Alfonso "Al" Constantino MV NEW DAYS at Longview, USA

contract. One of his main jobs, he says, is to make sure the crew's relationships are harmonious. "I organize birthday parties and join the crew to sing karaoke with the Magic Sing machine in the evenings."

Captain Alfonso asks, "Your dad is so strong. What is his secret?" "Yoga," I reply, "and he works in the forest." "Does he smoke?" "No, but he drinks . . . and his mind is clear—he doesn't worry." The captain nods.

He tells me his uncle was also a ship captain. Hearing his uncle's stories made him dream of becoming one too. He graduated from the Philippine Maritime Institute in 1978 and has been traveling the oceans since.

13 March '19, 11:45. All completed loading of logs. 1230—All completed lashing of logs. Ready for departure for Nanao, Japan. On deck 13,650 pieces to be unloaded in Japan. Underdeck, 24,709 pieces to be unloaded in Korea and China. Total Cargo 6,409,000 bf

6,409,000 board feet of wood
1,300 log trucks
200 acres of 40-year-old trees
the numbers stagger
the appetite
of the world's population
overwhelms

13 March '19. Chief Officer, Elisio I. Santillian (In charge of Loading/Unloading) Augustos Abroguena (3rd Officer In charge of Safety Equipment on board).

Dad and I talk with the officers as they steer the ship. Computers do most of the work, but the chief officer and third officer monitor them. After the long voyage with just each other, the men are happy to talk and answer our questions.

Augustos prefers log ships to other cargo ships because they take a long time to load and he can have shore leave. He shows us photos of him at a Trailblazers basketball game last fall. Basketball is big here on the ship. At lunch in the cafeteria, a taped game shows LeBron James still playing for the Cavaliers. At breakfast the next morning, the same game begins again.

The steel walls of the cafeteria create a time capsule. Sailors come in after completing their shifts. The same tasks, day after day, but done in different latitudes, longitudes, and weather. They watch the Cavaliers make the same plays on the same wooden floor. Long over, the live competition gives the sailors a still place, on a ship that is never still.

Another crewman, Rodrigo, becomes our guide on the ship. He is a philosophical man. "This is our life. This is our profession. We need to embrace it." "After nineteen days, we see the shore. I'm a little sad, a little happy, mixed together." "Sometimes seagulls are our companions here." "To prevent boredom we do maintenance."

About five thousand vessels a year follow the North Pacific Great Circle Route from the United States to Asia. Each must go through Unimak Pass, a strait in the Aleutian Islands of Alaska that is only nine nautical miles across. MV *NEW DAYS* will unload its logs in Japan, Korea, and China.

On the way back, sailors fill the ship's holds with seawater as

15 March '19. Weather Forecast. It was bad weather. A tropical storm all around the vicinity off Unimak Pass. Vessel had to maneuver and slow down speed.

ballast. I was disappointed to learn this; when teaching Ancient Egypt to sixth graders, I learned that mummified cats had been used as ballast on ships sailing back to England; the cats were then used as fertilizer. Even the ballast bricks used as paving stones in Boston are more interesting than seawater.

I ache to have been on MV *NEW DAYS* as it maneuvered this storm. I wonder at my desire to be on this ship. Is it the tangential energy and motion of strangers? The wildness of the ocean? The single-purpose nature of its voyage? I think of working in the woods and sitting on stumps. I like to be the observer, to notice and support. And I love the danger of the sea. The darkness of its depths calls to me. The summer Steve died, I floated for hours in the ocean in Hawaii, the waves taking over the movement of my body and muting the thoughts in my head.

In the engine room, next to safety posters hang two full-color posters: Jesus with his crown of thorns and the Virgin Mary with angels on each side of her veiled head.

17 March '19. Vessel encountering heavy seas in position L 48.45.1 N, X 138.21.9 W. Speed 8 Knots

25 March '19. Chief Engineer of NEW DAYS. Mr. Pilardo A Alama.

Captain Alfonso says the Holy Rosary each Sunday. "It is not compulsory. We have different religions: many are Catholic, some are Protestant, some are Seventh-day Adventist. We come here to work, but Sunday is the Lord's day. We need to give thanks. And it's good. We haven't encountered big storms or had a crew accident." I want to hear the Rosary, but we will not be on the ship on a Sunday.

It's the messman's first trip across the ocean. He misses his nine-year-old daughter back home. He went to culinary school in Manila and he makes pumpkin muffins for Dad and me.

26 March '19. Barbers Haircut at sea. Barber=Messman

The entry-level men, ordinary seamen, make $900 a month. Most send $700 home to their wives and use the rest to buy shoes for their families and "private food" to add variety to the ship's kitchen offerings.

After a dinner of grilled pork, rice, and salad, Dad and I head back to our rooms. We're both beat; it's just our third day in Japan and jet lag has caught up. My room has a window facing the back of the ship, a single bed, a desk, and a plastic-covered bench. I wash my face in the bathroom; one faucet pivots between the shower and the sink. I lie down and feel the six-hundred-foot ship gently roll through the sea. The movement eases me into a deep sleep.

When the sun rises, I sit outside on the deck, in a corner softened with a black rubber mat. Wind whips around the steel corners and snaps the flags, but I am protected. It has been just twenty-four hours, not twenty days, but I don't care. I write.

Later in the morning, a crew of sailors adjusts the lashing, hundred-pound chains that keep the deck logs in place. I take a video. Five men in padded orange or blue coveralls line up and heave a chain in unison. The first in line takes the slack and runs up over the stacked logs. The four other men continue the rhythmic pull until the chain stretches to the middle. They move to the next chain and repeat the demanding task. The Japan Sea is calm and blue under a clear sky. But the men must do this job twice a day, every day, no matter how rough the weather or waves.

We begin our approach to Maizuru, a port deep inside an inlet of the Japan Sea. We pass hundreds of small islands where wild cherry trees decorate the green hillsides with light pink smudges. Ken has chosen a glorious season for our trip.

As MV *NEW DAYS* draws toward the dock, I see Ken waving. I raise my arms in a victory sign. As we step from the ship, he says, "A dream come true."

In her twelfth-century Pillow Book, *Sei Shonagon lists Arima Hot Springs* as one of the top three hot springs. Ken has made reservations for us here and tells us we should bathe three times: before dinner, after dinner, and again in the morning prior to leaving. We are happy to oblige.

Yukatas, cotton kimonos, lie folded in the closets along with a belt and a small towel. Down the hall and up an elevator to the bath, we walk gently, not wanting to fall out of our sandals or cause our yukatas to gape. Blue curtains indicate the men's and red the women's. I wish Dad luck, take off my slippers, and enter the changing room. I place my yukata in a basket and walk naked through sliding doors into a large room with several pools. A sign next to one pool, filled with "gold water," claims its iron deposits are good for skin ailments and muscle pain. A pool with "silver water" contains radium and carbonate to soothe muscles and joints. As I step into the gold water pool, I am brought back to my first public bath.

1985, Fukui City, Fukui, Japan

The house I share with a fellow teacher does not have hot running water. To bathe we must fill the tub with cold water, turn a switch, and wait an hour. I soon discover the local public bath. Women sit on low stools in front of faucets lining the tile walls. They scrub with long, rough towels and pour water over their bodies to rinse the soap. Other women and children, already clean, soak in a large pool. When I walk in, there is a pause as they notice my foreignness. But unlike on the sidewalks where children follow me singing the ABCs and in shops where the clerks giggle and hide, these women welcome me. They scrub my back and are patient with my mediocre Japanese. During that year in Fukui, the bath is the place I go when my soul feels raw and alienated.

Dad comes back from the men's bath and groans with pleasure. His skin is pink with heat. Kimono-clad servers bring dinner to our room: twelve small dishes of salmon, tofu, seaweed salad, dried plums, thin slices of rich Kobe beef to dip quickly in a hot broth. Dad sits at a low table on tatami, his yukata loosely wrapped. An exuberant smile ripples the air around him as he holds a nasturtium with his chopsticks. Dad loves food. He loves new places. His joy makes him a great travel partner.

The next day we ride the train to Gero, a second hot springs town, one that Dad visited in 2007 as part of an Oregon State University Extension forestry tour. It's our first venture on public transportation without Ken, and Dad is nervous. I show him how to figure out which platform we need by comparing our tickets to the reader boards in the train station. But this involves Dad stopping, taking out his glasses, and looking back and forth while people bump into him from behind. I ask him to just follow, and he does. We get on the right trains and arrive at Gero in a cold rain.

After a morning soak, we hike up into the hills above the Hida River. A paved road turns into gravel, and we are in a managed stand of sugi and hinoki, two types of cedar native to Japan. We see a tree with a vine strangling it. Dad asks me whether I have my clippers. On our way back down the hill, we come across a small logging operation. The workers have stopped for lunch so we talk with them. A logger shows us how to tell the wood of the sugi from the hinoki—sugi has a reddish circle in the middle of the log; hinoki is pale throughout. Dad wants to know whether the machine sitting in the clearing will be used to load the logs onto a log truck. This stretches my vocabulary. I know how to say "carry," but "load"? I try and am rewarded with a smile and an answer.

Dad goes back to the hotel, and I keep walking. Paths follow both sides of the river through town. Pale cherry blossoms and bright pink blooms of an unfamiliar tree lie against a gray sky. The wind is cold and I wrap my red scarf around my head. In a driveway to the left, a bent old woman pushes a cart to and fro. I am glad my parents can still walk freely.

Being alone gives me time to let my mind wander. I start to realize that this account I'm writing about our farm and its produce is a way for me to create something from the trees. I'm not a forester like Dad, or an advocate for tree farmers like Steve, but I can tell a story about our land. This book is a gift from me to the land, to my family, to Dad, to Steve, to the people who work the land and love it. It is me making sense of the grounding the land has given me in my fifty-six years. A call to see the earth and what it gives.

The path has led me to a sign telling a local legend. I think about the Raven story as I begin to read.

The Utazuka Legend

Long, long ago, by the Hida River, there once lived two youths who sang beautifully. Tanosuke was the name of the lad living to the west of the river, while the boy who lived to the east was called Minosuke. Born with beautiful voices, these two young men delighted the villagers with their song. When the village leader began his search for a husband for his only daughter, Otami, the villagers spoke about the two youths as eligible candidates. The two young men developed tender feelings for the beautiful girl.

It was decided a singing contest would be held and the winner of this contest would become Otami's husband. The beautiful voices of the two young men gradually turned plaintive and ultimately took on a tragic air. At first light of dawn on the second day of the singing contest, the young men both died, having exhausted themselves through their singing. The villagers erected tombstones to appease the youths' departed souls.

Thereafter, Otami decided to live a chaste life, shaved her head and became a nun, and lived in seclusion in a hermitage at the top of Mt. Kobo.

Even today, every August, the towns along the Hida River hold a memorial service for the two youths. People living on opposite sides of the river compete in a singing contest.

I love legends, with their stories of tragedy and love grounded in light and darkness, elements of water and geography. This one is a little troublesome; Otami has no agency until she decides to shave her head, but still, I recount the story to the next few people we meet.

From Gero, we could just take the train back to Nagoya and, with another short train ride, be at a hotel near Camp Fuji, our next destination. But I've learned from Dad it is no fun to backtrack: "It's always good to be in new country. Always thought there would be better timber over the next ridge. I went to look many a time . . ." We take a circuitous course involving a train to Takayama and then a bus that winds through rugged snowy mountains and green valleys to Matsumoto, where instead of visiting the famous Black Castle dating from 1504, we spend thirty minutes trying to figure out how to work the coin lockers. We do manage to have time for a bowl of steaming buckwheat noodles before boarding the first of four more trains.

On the train to Takayama, Dad sits across from me on the almost empty car, looking out the window while mentally calculating the volume of wood in the cedar forests. On the last, a commuter train to Gotemba, he is squished between two teenagers on their phones, his body set in his "I'm in Zen mode, nothing can bother me" stance.

We are late to our hotel and miss the dinner hour. I've spotted a vending machine that sells beer and we make a dinner with my emergency stash of hazelnuts from the Portland airport, rice crackers from Gero, and chocolate-covered almonds from one of the day's train stations. We eat uninterrupted in our room. I am relieved to eat a simple meal and just be with Dad; no cross-cultural interactions needed. We sleep soundly.

At Camp Fuji, Dad stands with Sergeant Major Wyble next to a Quonset hut. When Dad was a first lieutenant here in 1957, one of his jobs was to oversee the building of Quonset huts for housing. Sergeant Wyble, a twenty-year veteran whose service includes four tours in Afghanistan and Iraq, has moved this last remaining Quonset hut next to the headquarters building. He plans to use it for displaying items from Camp Fuji's history. Mount Fuji, a few low clouds obscuring its peak, makes a dramatic background. Both men are flat on their feet, butts tucked in, shoulders back—a posture I was reminded to take countless times in my childhood. Dad tells Sergeant Major Wyble a story from 1957.

After I left Camp Fuji, I was shipped back to Okinawa to complete my two-year hitch in the Pacific. We were camped just off the beach and a terrible typhoon wiped out the whole camp. Our cots and tents were gone; all our clothes were wet and hanging to dry on anything still standing. I had to do something. Kadena Air Force Base was next door and I had done a little snooping there. They had materials for Quonset huts all stacked up, and I thought maybe we could borrow a few of those. I got a Marine Corps lowboy and a driver. We drove up to the gate, the guard threw me a highball, and I saluted him back. I went over to the forklift operator and asked him to put two Quonset huts on our lowboy. A week later two Quonset huts housed twenty-five marines each. As far as I know, no one knows about our "midnight reacquisition" even today. I mean, it was all the armed services. It's all ours indirectly, why not put it to use? And doing it on the sly makes it more fun, too.

Sergeant Wyble laughs, impressed. "We borrow a few engine parts now and then, but you borrowed a whole building!"

The Quonset hut turned museum is the only building left of Dad's Camp Fuji. The original base has moved across the road; the Japan Self-Defense Forces occupy the land where Dad spent his year. Mount Fuji still stands.

A few trains and a plane later, Dad and I arrive at Gimpo International Airport in Seoul. Pil Sun Park, a friend of Steve and Lou Jean's and a forestry professor at Seoul National University, is our local guide and will meet us here. On Pil Sun's first day of graduate school at the University of Washington, she met Steve, who was assigned to be her guide. They became fast friends and it is clear why; both are gregarious and have intense energy. Pil Sun is perfect for our quest. Not only is she a tree expert, but she loves history and is eager to share what she knows.

Dad, Pil Sun, and I stand at the Bell Pavilion at Gilsangsa, a Buddhist temple in Seoul. We lightly touch the red-painted wooden pillars and look up at the layered roof. Bright green, red, and orange paint decorates the ceiling, each square a mandala. It is midmorning and a chill rain is falling. The rain has kept most tourists away and the grounds are still. Tight pink azaleas bud under pine trees.

Dad and I have come to Seoul in search of a temple using wood from the Pacific Northwest, maybe from our land. The Weyerhaeuser spec sheet lists "temple log" as one of its highest-paying Douglas fir sorts. This use of our wood thrills me; flesh born from Cowlitz/Willoughby/Boone soil, tended by Dad and Steve, and now shaped by its new owners to make a sacred space. I want to touch the temple wood and listen to its stories. Pil Sun has contacted a few sawmills that import Douglas fir and found one that sells wood to temples. But recently they've stopped returning her calls. So we must conduct our own search. Because many of the buildings on the grounds of Gilsangsa have been remodeled or rebuilt, we hope we will find some Douglas fir from the Pacific Northwest here.

Gilsangsa is the home temple of Pil Sun's mother, Mrs. Cho. After inspecting the wood of the Bell Pavilion, Pil Sun says she wants to go to the main hall and pray to the Buddha there, the Buddha of Infinite Light. I ask whether I can come. Dad stays outside to look at more wood; the time it takes him to sit down and untie and remove his shoes deters him from entering. Pil Sun shows me how to pray: press hands together by the heart, bend, kneel, head to the floor, palms up, palms down, stand. Repeat three times facing the Buddha. I like using my whole body to pray. There is no question that I am doing something. It moves prayer out of my head and into my body, and the supplication seems to remain as I use my legs to walk out of the prayer hall.

We come out and ask Dad what he's discovered. He shakes his head. Temple builders prefer to use Korean red pine, but the Japanese colonization of Korea (1905–1945), World War II, and the Korean War devastated the local forests. Red pine large enough for temples is hard to find. Douglas fir is in the pine family, and once the bark and foliage are removed it is difficult to distinguish the wood. We look again at the walls of the main hall: red pine? Douglas fir? I'm with two tree experts and they don't know. But I imagine that Gilsangsa's main hall was built from wood we logged in 2011, just above the spring where I talked with Esther Willoughby. With that thought, I put my hand to the wall and listen.

On the way to the next temple, Pil Sun tells a story that shocks. I find newspaper articles about the incident and compose a found poem:

Sungnyemun Gate
from the *Korea Herald*, January 3, 2014 / *Korea Times*, January 20, 2014

Sungnyemun Gate
National Treasure No. 1
one of the four gates
that surrounded Seoul during the Joseon Dynasty (1392–1910).

A parliamentary audit in October 2013
revealed
flaking paint
wooden columns split down the middle.

Speculations have grown
craftsmen allegedly brought red pine
imported from Russia and other regions,
instead of using homegrown red pine

A Chungbuk National University professor
a renowned expert of wood anatomy
tasked to look into allegations
found dead over the weekend in an apparent suicide hanging.

After hearing this I do not mind not knowing whether the wood we see in the temples is "ours" or not.

At the second temple, I sit in an empty hall, surrounded by wood—heated plank floor, soft brown pillars, curved arches. Three golden Buddhas watch over me. I cannot read the prayer books on the pillows, so I sketch the shape of a lotus behind one of the statues. Silence seeps into me. Alone for a moment, I breathe deeply, wood-imbued air fills my lungs.

Pil Sun and Mrs. Cho have invited us to Gyeongju, the capital of the Silla Kingdom, 57 BC to AD 935. Buddhism was introduced to Korea from China during this time, and we've come to visit temples and also aboveground tombs that still protect royal members of the Silla Court. Laborers layered river rocks up to heights of 12.7 meters. Below the stones a wooden room houses a coffin and gold items for the afterlife: antler-shaped crowns, intricate necklaces, earrings, statues of Buddha. Nearby we examine a Silla era fourteen-sided die found in 1974. Each side gives instructions to guests to follow during parties. Pil Sun translates: "Drink three glasses of liquor at one time; have one's nose struck by many companions; sing and drink all by yourself; stay still even when your face is being tickled; recite a poem."

The next day is Dad's eighty-sixth birthday. After our breakfast, Mrs. Cho orders a piece of cake and places a candle in it. Dad's deep-set eyes smile as he makes a dramatic blow on the flame. His back is straight and his mind clear, but as he claps, his gnarled hands belie his age. At Gyeongju Bulguksa, a temple completed in AD 774, visitors can ring the fifteen-foot-long iron bell. Dad stands between two red pillars (red pine? Douglas fir?), pulls back a six-foot-long wooden beam, and makes the bell echo into the valley. I hope I am ringing a bell in a foreign country on my eighty-sixth birthday.

Inside the temple, I pray next to Mrs. Cho, using my newfound skills. Through Pil Sun, she asks, "What is your religion?" "Nothing, but I am interested in all religions." A big smile, a thank-you in English. I smile back. I can say only "the weather is nice today" in Korean. Our words do not connect, but we are united by three bows, three foreheads to the worn smoothness of the wooden planks.

PART FIVE

Heartwood

Heartwood is the central, supporting pillar of the tree. Although dead, it will not decay or lose strength while the outer layers are intact. A composite of hollow, needlelike cellulose fibers bound together by a chemical glue called lignin, it is in many ways as strong as steel. Set vertically, a one-by-two-inch cross section that is twelve inches long can support twenty tons.

This year's harvest is

    2.5 million years from the placement of soil by glaciers

    ca. 5,000–10,000 years from the arrival of Cowlitz peoples to the valley

    400 years since the trees cut in the early 1900s began to grow, most likely after a wildfire

    198 years from the marriage of Thas-e-muth to Simon Plamondon

    192 years from the death of Thas-e-muth

    190 years from the flu that killed 90 percent of the Cowlitz peoples

    180 years from the founding of Saint Francis Xavier Mission

    163 years from the refusal by the Cowlitz to sign a treaty and move to a proposed reservation

156 years from the Homestead Act, which allowed the Willoughbys to own land

129 years from Washington gaining statehood

126 years from the incorporation of Toledo

110 years from the first clear-cut on our land

105 years from the commissioning of the Pacific Highway (old 99)

98 years from the opening of Wayside School on land deeded to the district by Elmer Boone

88 years from the closing of Wayside School

58 years from the building of Interstate 5, bypassing Toledo

38 years from the explosion of Mount Saint Helens. Ash lies buried under thirty-eight years of duff

33 years from my first move to a city, Fukui, Japan

18 years from federal recognition of the Cowlitz Tribe

15 years from my move back to the West Coast, back to the land

4 years from Steve's passing

Tomorrow, Four Seasons Forestry Services will start planting our clear-
cut. Today, Dad and I are off to gather 6,600 seedlings. We drive the
truck and trailer through the old timber towns of Tenino, Rainier,
Yelm, and McKenna to Roy, Washington. Dad points out a cemetery.
"A few old loggers are buried up on that hill. This whole area was easy
pickings for early logging. The flat land was cleared and turned into
farms."

In Roy, Silvaseed processes conifer seeds for large timber companies and grows seedlings for small landowners like us. After we load
our bags of western white pine into the trailer, I tour the warehouse.
Stacked inside a walk-in freezer are containers with labels for Den-
mark, Germany, the Netherlands, the United Kingdom, and Swit-
zerland. They are full of seeds. I learn that each August, Silvaseed sets

up seed stations all over Washington and people bring cones from different elevations and soil types. Silvaseed dries the cones, removes the chaff, discards seeds without embryos, and freezes finished batches for five, ten, twenty years.

I thought the planting was the beginning of our forest, but a whole new chapter has opened up, with more people, more expertise, more science. Where is the beginning?

Pine purchased, we head to Tumwater, where Chris Whitson of Port Blakely is storing western red cedar and Douglas fir for us. We cover the boxes and bags with a blue tarp and head back down I-5, the start of a new forest in our trailer. My belly is full of anticipation. I've been waiting for the planting. I want to talk with the men on the crew, watch and listen to the process, be part of the job, be part of the bare land growing again.

Tim from Four Seasons calls. His mother has just passed unexpectedly at sixty-nine. They will plant Tuesday and Wednesday, have the service on Thursday, finish up on Friday.

In the evening, I make oatmeal and raisin cookies for the crew. Maybe at lunchtime they will sit outside so I can join and ask questions.

At 7:10, a white Four Seasons truck pulls up to the house. Dad and I leave our unfinished breakfast on the table and drive out to the cutover with them.

The crew—Guillermo, Abraham, and brothers Don and Tim—put on their gear, gather equipment and seedlings of western white pine, western red cedar, and Douglas fir, three species chosen to outwit root rot and global warming. The pine are one-year "plugs," so small they disappear into the deep pockets of the bags and are hard to spot after planting. The bushy tips of the fir and cedar bush out up to the planters' shoulders. Tim, Don, and Abraham work side by side, nine feet apart and pacing nine feet between each planting.

Scrape, scrape, scrape
shovel metal against river rock
laid down by glaciers thousands of years before

tap, tap, tap
boot heels on blade steps
push-pull, push-pull, push-pull
right hand on the handles
the ground opens up
left hands reach back to grab a seedling
bodies bend to gently place
the tree in the hole
men stand up
pull the shovel from the dirt
tamp down the soil around the planting
and walk another nine feet.

Lou Jean, Dad, and I tube the cedar as they are planted. Deer can munch the thriving green fronds into nubs overnight. We pull the innermost tube from yellow bundles of ten and weave a bamboo post through the mesh, bend to place the tube over the seedling, then push the post deep into the soil. We stand up, look nine feet ahead, and wait for the next seedling to come into focus against the dark soil and green ferns. Rocks obstruct my next post; I move it two inches to the right and it slides right down. Izzy finds a deer bone and happily chomps away. Dad stops tubing and prepares a backpack sprayer with Plantskydd, a deer deterrent made of pig or cow blood mixed with water. We will watch to see which cedar flourish best—those tubed or those treated with Plantskydd.

Tim is planting next to me.

I weave another tube. "You know, the last time you guys planted for us, it was right after Steve passed."

Tim taps his shovel with his boot. "Yes, I remember him being sick."

I place the tube over the tree. "It's good to be planting after someone dies."

I've had almost five years to gain distance from Steve's death. Tim's mother's service is in two days, and he looks at me quizzically but graciously replies, "Mmm, the cycle of life."

At lunch, rain has started to fall and the planting crew sits in the truck. I can't interrupt their meal with my questions, but the door is open and I hand them the cookies. Surprised, Abraham takes them; Dad and I head to the house, where I make grilled cheese sandwiches with Mom's dilly beans.

I decide to follow Guillermo around with my notebook as he works.

I ask whether planting has changed any.

Shovel handles used to be wood—lighter, but broke too easy.

Tree bags have shoulder straps now, not just padded waistbands—more comfortable.

Companies want more area scarified.

Weyerhaeuser has changed planting spacing from 11.5' × 11.5' to 12' × 12'.

No more broadcast burns.

I ask Guillermo what he thinks about not burning at all, leaving more slash on the ground. He replies, "That's crap." He says it would decrease the number of trees a worker could plant each day from 1,100 to 800 and so raise the cost of each from thirty to thirty-eight cents.

I work five hours. My back hurts, my hands are sore, and my mind is starting to go numb.

Dad is still tubing. Embarrassed to be outworked by my octogenarian father, I keep going. But after another hour, I head for home down the gravel road. I say hello to the saw-whet owl's home, to the branch that held the barred owl, and crawl between the sheets. At the annual meeting of the Lewis County Farm Forestry Association that night, Dad's legs cramp up and he is nauseated. He tells Mom he is going to work 10 percent less.

The first day of planting is a full moon and Mom's eighty-fifth birthday. We've been celebrating with liver pâté, cheese, and yellow roses. A friend has given her a book about Georgia O'Keeffe. "I'm relieved it's not the book by Rumi she and I have been discussing," Mom tells me. I ask why. "My relationship with Georgia O'Keeffe is timeless, not a burden." I want to know more. "At eighty-five, I can't keep up with a lot of things. I'm shedding chores, friends. Georgia O'Keeffe is not one of them."

The next day, Tim pauses as he packs his planting bag with cedar seedlings. "We had some cookies left over. My son and granddaughter are here from North Dakota for Mom's service. I gave the extras to her—she loves cookies." It's the first conversation he's started and I feel like I've been given a present.

I ask Tim what makes a good planter. "Well, we try to put just one tree in each hole—some guys try to give the bonus. And I try to get the green part up, the roots down." Don and Abraham laugh. It's nice of them to indulge my questions.

Guillermo was born in Michoacán, Mexico, and came to the United States in the 1980s. He's worked in reforestation since then, planting his first trees in the blast zone of Mount Saint Helens. This weekend he will plant trees on his own forty acres of land, just across the Cowlitz River from our farm.

I ask Tim and Don when they first planted trees. "Dad got us out pretty little," Tim says and indicates a height at about his waist.

"Did you get paid?"

"I don't think so."

"We used to earn a dollar an hour. If we wanted a raise, we had to tell Dad why we were worth more."

"At least you got something."

After our second planting day, the trailer stands almost emptied. I will not be at the farm for the final day. I shake hands with Tim, Guillermo, Don, and Abraham. "I really enjoyed working with you all. Can I call if I have more questions?"

When I return to the farm that weekend, I ask Mom whether she's seen the obituary for Tim and Don's mother in the *Chronicle*. I read her my favorite line: "Caroline enjoyed all activities the kids were involved in and was a strong supporter of Bearcat Wrestling, where at times it could be risky sitting next to her." I notice that she graduated from Toledo High about the time Lou Jean's brothers did. When Lou Jean reads it, she exclaims, "Her mother was my piano teacher." We'll have to tell Tim and Don the next time we see them.

A few days later, I walk out to the clear-cut. No shovel sounds, no voices, just two geese crying after each other along the ridge. I sit on a cedar stump and take out my notebook and listen. In the stillness, growth has begun. Surrounding my stump, six seedlings push their roots into the soil, stretching their spindly branches and needles to pull in the sun. Stumps sequester carbon dioxide and provide shade from the August sun. Deer droppings and logging slash decompose to form mulch, and beetles and earthworms tunnel, adding oxygen to the soil. Dad has named the cutover "Two-Snag." He knows that Steve will appear as a raven or a red-tailed hawk and perch on the two long-dead trees we've left.

I look out at the expanse of twelve acres: empty, but not empty. I am reminded of my nine-year-old self ducking my friend's head. I feel the absence of the tree that was cut from the stump that supports me. I am glad we're saving the old trees in Gemini Grove.

The cutover is near the house; it will be easy to get here and tend these seedlings. The trees in the clear-cut need me, need me to mulch them, to adjust the tubes protecting them from deer, to pull Scotch broom, cut Himalayan blackberries and honeysuckle. Their need satisfies a deep want in me. I wonder at the roots of this emotion, whether it helped propel hunter-gatherers to plant and domesticate. Why aren't humans satisfied with what the land gave them? Why ask more of the forests, the prairies, and the seas?

Mount Rainier is out in its full glory, double peaked and brilliant white. Snow fills the clear-cut patches in the Cascade foothills.

Acknowledgments

This book has risen from grief, from the ashes of my brother's too-early death. I hope the story I tell about the land he loved serves to uphold his memory.

Thank you to Mom, Dad, and Lou Jean Clark for allowing me to share parts of your stories. I am honored to be trusted with them.

I have talked about the ideas in these pages with many people. Those conversations helped sharpen my thoughts and often gave way to new understandings. In particular, thank you to Laura Neitzel, Jennifer Newton, Aron Wagner, Luci Kegley, Martha Zornow, and Bob Herr. Laura's questions helped shape the form of several chapters, and she is the muse behind the title. My sister, Julie, has been a cheerleader from the beginning.

Many of these chapters got their start in writing workshops around Portland. Tom Kizzia at Fishtrap, Ann Staley with Creative Arts Community, Kim Stafford at Lewis and Clark, Jennifer Denrow of Literary Arts, and Mark Cunningham with Atelier 26 Books all provided inspiration and encouragement. Poets I met in a group started by Shelly Cato and Melody Wilson have improved the book's poems.

Meeting Mark was most fortuitous. His faith in this project gave me the courage to keep writing. Because of him I think of myself as a writer. His careful readings helped me see larger themes in my story, and his ear for language made my sentences stronger.

And so many people helped me gather information for the research sections of the book. I'd like to thank Roy Wilson and Michael Hubbs. Both gave generously of their time and were patient with my many queries about the Cowlitz Tribe, to which they both belong.

Research done by the Toledo Historical Society gave me a strong basis from which to begin imagining the lives of early settlers on our land, and the librarians at the Washington State Library helped me find documents to flesh out their lives. Thank you also to Julie McDonald Zander for reviewing the chapter about Toledo history and to the Buswell family for sharing their insights about early logging in the area. The Arbor Day Foundation penned the descriptions of tree layers I've used as section introductions.

I could not have made the trip to Japan without the persistence of Mike Warjone of Port Blakely Tree Farms. He helped convince NYK Shipping that Dad and I would be safe guests and recruited Ken Shioda to be our guide in Japan. Ken not only shepherded us between the ports and sawmills but also took us to fabulous ramen and soba shops. Special thanks also to Captain Shoko Sakai of NYK Shipping, Takanao Okazaki of Chugoku Lumber Company, and Sergeant Major Wyble at Camp Fuji.

The crew of MV *NEW DAYS* made Dad and me feel instantly at home as we shared their space from Nanao to Maizuru. Thank you to Captain Alfonso Constantino for taking photographs of the cross-Pacific voyage so we could live it vicariously.

Pil Sun Park and her mother, Mrs. Cho, were our guides in Korea. We were in awe of Pil Sun's skill in Seoul traffic and her deep knowledge of Korean history. Mrs. Cho's snack bag with chocolate and fruit kept us satisfied on our travels to the countryside.

I am thankful to all the people who put up with my incessant questions: Peter (Jr.) and Peter (Sr.) Mahnke, our loggers, taught me so much. And many thanks to Peter Sr. for allowing me to include his poem. The tree-planting crew with Four Seasons also indulged my questions and gave me new insight into their profession. To the wonderful people at Reichert Shake and Fencing, thank you for allowing me to tour the mill.

Thank you to all tree farmers for your work protecting water, air, and wildlife all while providing jobs and wood to our communities.

Thank you to the publishing team at Oregon State University: Tom Booth, Marty Brown, Kim Hogeland, and Micki Reaman, and to copy editor Laurel Anderton. I can't imagine all the details that go into making a book out of a manuscript, and getting that book into readers' hands.

And finally, Tom Barbara, thank you. This book exists because you are in my life. You provide a rooted place where I can be still and create.

Author's note: Any mistakes are mine alone. Names may have been changed or omitted to protect the living.